W9-CCD-167

Roadmap
to the
e-Factory

ALEX N. BEAVERS, JR.

AUERBACH

Boca Raton   London   New York   Washington, D.C.

Library of Congress Cataloging-in-Publication Data

Beavers, Jr., Alex N.
    Roadmap to the e-factory/ Alex N. Beavers, Jr.
        p. cm.
    Includes bibliographical references and index.
    ISBN 0-8493-0099-1 (alk. paper)
    1. Computer integrated manufacturing systems.

TS155.63 .B43 2001
670′.285—dc21
                                                                    00-065054

This book contains information obtained from authentic and highly regarded sources. Reprinted material is quoted with permission, and sources are indicated. A wide variety of references are listed. Reasonable efforts have been made to publish reliable data and information, but the author and the publisher cannot assume responsibility for the validity of all materials or for the consequences of their use.

Neither this book nor any part may be reproduced or transmitted in any form or by any means, electronic or mechanical, including photocopying, microfilming, and recording, or by any information storage or retrieval system, without prior permission in writing from the publisher.

The consent of CRC Press LLC does not extend to copying for general distribution, for promotion, for creating new works, or for resale. Specific permission must be obtained in writing from CRC Press LLC for such copying.

Direct all inquiries to CRC Press LLC, 2000 N.W. Corporate Blvd., Boca Raton, Florida 33431.

Trademark Notice: Product or corporate names may be trademarks or registered trademarks, and are used only for identification and explanation, without intent to infringe.

Visit the Auerbach Publications Web site at auerbach-publications.com

Roadmap
to the
e-Factory

Dedication

To Linda, who always believed in me,
and to Chris, Alisa, Amanda, and Robin,
who always inspired me.

Contents

Purpose of This Book

Having been in the business of manufacturing as an owner, operator, and consultant for the bulk of my career, I have seen the ebb and flow of technology trends in manufacturing similar to an ocean sending waves to shore. For the last 30 years, there has been wave after wave of first promises, then expectations, and finally the harsh realities about the "breakthroughs" or "revolutions" in manufacturing that would transform overnight the way that companies create value from purchased goods and services. However, because of technologies related to the Internet, there are changes taking place today throughout the economy, across geographies, and within the thought processes of industry leaders that are creating such fundamental changes in how companies decide where they add value and how they satisfy their customers' needs that the wave analogy does not apply anymore. The phenomenon is more like a melting of the polar ice caps and a flooding of the continents. Along with this epic change comes opportunity and risk. The opportunity is to rethink how a manufacturing enterprise operates. The risk is that there are few clear-case examples of how to do this successfully.

The purpose of this book is to address the opportunity and risk that come with creating the e-factory. This is accomplished by (1) providing insight into the transformations occurring in manufacturing because of e-business; (2) providing an organized approach to analyzing the technology and process needs of a manufacturing organization due to these transformations; (3) identifying the technology solutions necessary to respond to the needs; and (4) providing a methodology for building a roadmap by which a company can navigate to successful solutions. This book is designed to be used by both executives of manufacturing companies who are beginning to plan for the transformation of their operations to e-factories and by university students in graduate school or in advanced industrial management classes who are preparing for the new Internet-based economy.

Alex N. Beavers, Jr., Ph.D.

Foreword

The dawn of a new century brings with it the convergence of technologies, business practices, and management principles that are creating major changes in the way companies in most industries operate. For manufacturing companies, these changes are often taking the form of major transformations within their enterprise and throughout the supply chains within which they operate. The managers of these companies are faced with many major decisions about what technologies and process changes to implement, how to implement them, and how fast they should implement them. The primary objective of this book is to describe a methodology for performing the analysis that leads to the best decisions about these questions.

In addition to a methodology, included in this book is insight into the nature and causes of the trends that are leading to this convergence and their implications for the future in the manufacturing environment. The purpose of these insights is to provide a frame of reference for the decision-makers who read this book within which they can think about future changes and management issues.

As with most popular trends that suddenly erupt on the scene in fashion, culture, or economics, there have been a variety of provocative terms coined for use by the general press about things related to the Internet. Bowing to this tradition, the manufacturing environment that is evolving as a result of the changes being created by e-business is termed the e-factory herein. The e-factory represents the manufacturing environment at the dawn of the 21st century that is facilitated by computer and communications technology in the form of embedded logic chips, personal computers, client/servers, and all other assorted equipment and software that are accelerating business processes and shortening activity cycles.

The changes taking place in manufacturing because of e-business trends are dramatic and provocative. They are dramatic because they reflect the need for a completely different way to think about a factory. A factory is becoming a node in an e-business network, a physical and electronic link in an electronic supply chain, and a producer of information and services as well as material products. The changes are provocative because they are causing entire industries to be restructured and disintermediated in terms of ownership, workflow, material flow, and information flow. Vertical integration and horizontal integration are business terms left over from the Paleozoic period of management thinking. Now, thanks to e-business, the discussion is in terms of Web integration and value chain integration. Business enterprises are now becoming virtual portfolios of business processes and supply chain webs.

The e-factory is a key part of this evolution because without dramatic and provocative changes in the factory, the threshold for change in the economy would be set very low. The purpose of this book is to provide guidance to those who are on the cutting edge of having to plan for and implement the technology, business processes, and decision-making practices necessary to succeed in the new e-business environment.

Chapter 1

Introduction to Business Cyberspace

Purpose of This Chapter

There has been much said and written over the last ten years about how business will be or is being conducted in cyberspace. While this "hyping" of the business utilization of the Internet has been aimed at promoting the cutting-edge aspects of companies offering products that are necessary for the Internet to continue growing, it has been a harbinger of great change in how business-to-consumer (B-to-C), business-to-business (B-to-B), and business-to-employee (B-to-E) will be conducted in the future. Although the nature and use of this hype has been changing yearly, there is a fundamental change in the structure of how business is being transacted due to the advent of the technologies related to the Internet.

The purpose of this chapter is to introduce several concepts and structural models that are key to describing what is happening in business cyberspace. The concepts of e-business and the e-factory and models that describe the key elements of each are introduced in this chapter. These models and concepts are used throughout the remainder of this book.

What Is e-Business?

One of the terms that was long ago overworked but which is still useful is "e-business." The definition of e-business includes everything from the electronic facilitation of business transactions using the Internet to a whole new form of commerce based on "outside the box" strategic thinking about where companies add value in their global supply chains. As a minimum, e-business is characterized by three adjectives: on-line, real-time, and interactive. On-line means that individuals and companies have computers that are actively connected to a network for a considerable portion of their decisions and transactions. Real-time means those individuals and companies are expecting answers to their inquiries or their communications within seconds. Interactive means that individuals and companies are conducting extended conversations, negotiations, or transactions while they are on-line.

The state of being on-line today is usually defined to describe the fact that people can electronically interchange with each other any form (e.g., voice, data, images, video, audio) of information using a broad range of communication technologies ranging from telephone lines, to cable TV lines, cellular phones, satellite dishes, and dedicated wire lines. However, the communicating entities today are more than people-to-people in chat rooms; they include people-to-machines (e.g., Web sites) and machine-to-machines (e.g., drive-through tollbooths, factory floor control systems, climate control systems). It is important to keep in mind that in the world of e-business, human communication is only a part of the overall web of information flow. Much of the celebration of e-business alludes to the fact that the spoken or typed word of the consumer is the controlling factor in the flow of commerce; the reality is that the spoken or typed word only sets off a complex set of electronic and human actions and interactions that eventually result in a delivered good or service.

The complex set of actions and interactions performed due to all these human and machine resources being on-line are also inherently and relatively real-time. The definition of real-time is a bit imprecise, but in simple terms, it means "really really fast" and certainly within just a few seconds. By harnessing the technology that allows large amounts of information to be accessed and distributed in seconds, by making it easy and inexpensive to be on-line, entire shopping / searching / decision-making / shipping-date-commitment / final-purchase-decisions can be performed during an on-line session. Whether this e-business "cycle time" is measured in minutes, seconds, or milliseconds, when compared to the business cycle times of the pre-e-business era, it is real-time.

Given shrinking cycle times and easy on-line access, more and more of the activities revolving around e-business are interactive. These interactive activities can take the form of people-to-people, people-to-machine, or machine-to-machine. They create a real-time thinking and decision-making process that has tremendous implications for the future. It is as if e-business has made it possible for the same impulses that retailing marketers of celebrity magazines and snack foods exploit at the checkout aisles in supermarkets to be exploited in every other aspect of personal or business decision-making. While there may be social and cultural implications of this trend, there are clearly major implications as to how businesses will operate in the future.

e-Business will cause many significant changes in virtually every aspect of our professional and social lives. However, e-business-based actions and transactions must add value or satisfy a need; otherwise, nonelectronic media will suffice. Over 20 years ago, the technology trend that preceded personal computers was office automation (does anyone remember Wang word processors?).

The claim then was that the "paperless office" would transform our economy overnight. Ever since, the amount of paper consumed and the number of paper copiers and printers has grown steadily. The same could be true for e-business; it was just a couple of years ago that some people were forecasting that the era of the retail store would be replaced by e-tailing. In fact, as revenue from on-line sales has increased dramatically, it still represents only a few percentage points of total retail sales while there is a renewed growth in new retail space, which is often bigger and grander or smaller and trendier. e-Business is causing significant change, but it will only replace those activities where more value can be created via electronic means.

Another key cultural change that e-business is stimulating is in the area of commercial relationships (relationships between the consumer and retailer or the buyer and seller). e-Business is creating commercial relationships that are reconfigurable and that are not dependent on loyalty, geography, culture, custom, or marketing message, but that are dependent on price and performance. One of the most important initial e-business trends that lead to the creation of many Internet companies was stimulated by the promise of finding low prices for products and services on the Internet. New on-line services allow a buyer is to shop for the best price for products and services by searching the Web rather than jumping in the car and driving all over town to visit many different stores or stopping at the nearest superwarehouse store. While this trend is emerging because of the fact that virtually every commercial enterprise has established a Web site to support on-line purchasing, other emerging trends indicate that on-line buying is not the price-killing dream that some expect. Studies are now showing that on-line buyers get impatient after a few clicks and either give up the search, or give up and buy a product from the first Web site they hit that has something like the product they want. It is also indicating that there is emerging a loyalty to Web sites that offer convenience in shopping as well as price. In fact, studies have also indicated that the average price of an item purchased on the Web is actually higher than what could be found from alternative non-e-business channels. In addition, there are new on-line services emerging that are making the value of an on-line relationship go beyond just shopping for price.

It is becoming clearer that one of the fundamental trends of e-business that will likely be sustainable is that business relationships are changing because they are on-line. Customers, suppliers, internal operations, and even the government are on-line to communicate, exchange information, perform transactions, and provide status information. An on-line relationship includes contact with different parts of internal operations, suppliers, and related third parties. An on-line relationship must manage material flow, product information flow, and transaction flow.

Customers Are On-line

There are at least two types of customers to consider. One type of customer is the consumer, the end user of most products and services. The other type of customer is the business customer who is acting on behalf of their employer to buy goods and services for their company. The business customer can be the procurement professional, the buyer from the purchasing department, an engineer working on a product design project, a maintenance worker looking for replacement parts, or an office worker looking for office supplies. While there are very different technologies and business processes evolving to serve both types of customers, there is a merging of the expectations that customers have and of the needs that they want fulfilled.

Consumers are on-line to browse for information and to satisfy their curiosity about social and business topics. They are on-line to get updated on relevant current events such as news headlines, stock prices, and to obtain information on where certain services or products can be purchased.

Consumers are on-line to find solutions to their problems, to seek help in defining their problem, and to define a solution for a problem. They are on-line to select the best solution that is best in terms of configuration, functionality, price, style, or availability.

Consumers are on-line to buy solutions, to place an order, to get confirmation on delivery and installation. They expect to receive more information on-line than they have ever received from a retail store or catalog purchase.

Consumers are on-line to get information and advice about their order. The information includes instructions about preparing for delivery or installation, the status of the delivery of the order, including the geographical location in the transportation system.

Consumers are on-line to receive new services. They have been conditioned by advertising and bombastic news reports to expect the introduction of new, wonderful, amazing services frequently and with blinding repetition.

Consumers are on-line to express opinions about social issues, product issues, religious issues, cultural issues, and political issues. The concept of the telephone survey pole has now been extended to include the Web site survey or e-mail survey. Reviews of restaurants, movies, political decisions, even management decisions, can now be found on a myriad of personal Web sites.

Consumers are also on-line to buy or receive entertainment. However, the entertainment industry has been somewhat disappointed in the course that this trend is taking. Consumers have been so conditioned by the early offerings of free information and entertainment (music, literature, art, etc.) that the growth of retail sales of entertainment has been less than expected. In addition, the sale of entertainment that requires high bandwidth for delivery, which is primarily video, has been slow in taking off because the availability of high bandwidth transmission to consumers has been low and slow in coming. Of course, it does take time to rewire the earth with high bandwidth optical-fiber cable or to put a satellite dish in every home.

Business customers are on-line for many of the same reasons listed above, but also for a variety of other reasons that pertain to the transactions between companies.

Business customers are on-line to find and negotiate with suppliers. Finding suppliers often comes about through research using search engines and keywords. Negotiating with suppliers will involve many of the same steps that are traditional to the procurement process: request for qualifications, request for proposals, proposal submission, and proposal evaluation.

Business customers are on-line to execute transactions in accordance with an existing contractual relationship. The transactions could be through a special Web page that the supplier has established for the customer's company that has built all the appropriate commercial terms and conditions such as price, delivery, warranty, and payment. Or the transactions could be through a catalog, or through a distributor's Web site that provides a more real-time service than catalog buying.

Business customers are on-line to report to suppliers the quality of their products and services. Companies are now beginning to report on the level of performance of their suppliers via special Web sites. This provides for real-time access to this information for the supplier as well as a corporatewide sharing of information by the customer.

Business customer relationships will be people driven. People maintaining customer relationships will use e-business tools as a facilitator and accelerator of customer activities, will have different skill sets and training, will be e-business enabled, and will be a key part of the total process.

Suppliers Are On-line

Suppliers are on-line with their customers and other suppliers to collaborate on product development. This is a best practice in supply chain management because collaboration can lead to lower cost products, higher quality products, and faster time to market. By being on-line, suppliers can provide existing product performance data, requirements for new products and services, assembly process data, and field data on installation or service issues. Suppliers are on-line to share data about capacity and delivery of new products and the cost drivers for new products.

Suppliers are on-line to collaborate on demand management. The objectives for this type of collaboration are to achieve shorter order fulfillment cycle times and to determine the best mix of build-to-inventory and build-to-order tactics.

Suppliers are on-line to collaborate on designing an integrated process for ordering and scheduling materials and services, to collaborate on developing decision criteria for key performance parameters (e.g., stocking levels, commitment levels, service levels, and planning revisions), and to collaborate on pricing policies.

Suppliers are on-line to collaborate on procurement and shipping policies in order to achieve more accurate and faster order fulfillment, and for lower cost operations.

Suppliers are on-line to satisfy e-business driven, supply chain planning and order commitment needs. This includes sharing information about capacities and inventory availability and jointly planning for future orders and capacity utilization.

Suppliers are on-line to implement and control the external business process that must exist between a supplier and customer who have a strategic or tactical partnership. This external business process environment includes the workflow for decision-making about new or existing needs, strategic information about available components or services, or ideas, other intellectual property, and controlled workflow for materials management, installation, and systems integration planning

Internal Operations Are On-line

Internal operations are on-line to respond with product and service information to customers and suppliers. For example, new product development concepts must include the capability to support rapid order configuration by non-expert customers who will be ordering the products over the Internet. Also, this implies that the order fulfillment process must be designed to be build-to-order and to execute transactions directly with customers, suppliers, and financial institutions.

Internal operations are on-line to share data internally. Internal sharing of data is necessary in the e-factory because all the control systems and operating principles are designed to operate in real-time. Real-time systems need real-time data. Thus, sharing data internally is critical to the e-factory.

The internal data that needs to be shared in the e-factory includes the throughput, yield, and quality of the order fulfillment process, the impact of new product concepts on field and factory operations, the need for new product concepts from the experiences and insight of the field and factory, and the operational readiness of all critical activities in the field or factory.

Internal operations are on-line to share data externally. Sharing data externally means sharing it with on-line partners that may be suppliers or customers. The data to be shared with customers depends on the nature of the contractual relationship between the company and the customer. Similarly, the data to be shared with suppliers depends on the nature of the contractual relationship between the company and the supplier. This need for data sharing is creating legal and ethical issues around the handling and ownership of intellectual property and the issue of collaboration on commercial terms such as pricing and warrantees.

Government Is On-line

In many ways, the federal government has been a leader in getting on-line with the electorate. This trend is growing at the federal level and is picking up speed at the state and local levels as well.

The government is going on-line to support tax processing and tax preparation by experts and non-experts. The government is going on-line to provide regulatory services. The government is also going on-line to provide educational and advisory services.

The interaction between citizens and government is rapidly becoming an on-line experience. Tax returns can now be submitted on-line. Some jurisdictions are experimenting with on-line registrations of various types and even on-line voting.

The military is on-line for both communication with the external world and for internal operations. In fact, the original funding for all the technology that is now the backbone of the Internet came from the Advanced Research Projects Office (ARPA) of the Department of Defense. The original Internet was ARPANET, a network of government and academic research computers that were linked to speed up the communication between private and public research centers.

An Enterprise Process Model

One of the results of the emergence of e-business is that companies have to rethink where they add value in a supply chain, how they do that profitably, and how they should transform themselves. To describe how these issues can be addressed, a few simple models of business "enterprises" are presented in the following.

The full impact of e-business on the manufacturing enterprise is best understood when a business process model is used. In reality, an enterprise consists of only a few key business processes. Business processes that are typical in a manufacturing enterprise include those of product realization, order fulfillment, order capture, customer service, and administrative support. This business process perspective of the enterprise is useful because it is inherently activity based and reflects the actual flow of information, work, and materials through the enterprise.

A business process consists of objects performing activities all for the purpose of delivering a good or service. The term "object" is used because the entity performing an activity can be human, machine, or data. Activities are actions or decisions being performed by an object. Activities result in a raw material or service being converted into a final good or service or in some value being added to a work in progress. Activities also have a cost that results from consuming resources and from consuming time. The cost of activities can be measured in currency. This simple set of concepts permits any aspect of a business enterprise to be modeled, described, controlled, and improved by dealing with a simple set of business process elements. All business processes are inherently a cross-functional and cross-organizational unit. A business process reflects the actual flow of work and material through an organization.

The diagram in Exhibit 1.1 illustrates a few of the key processes that exist in most manufacturing companies: the product realization process (PRP), the order fulfillment process (OFP), and the Order Management Process (OMP). There are other key processes that are important as well in that they provide support to the three key operational processes.

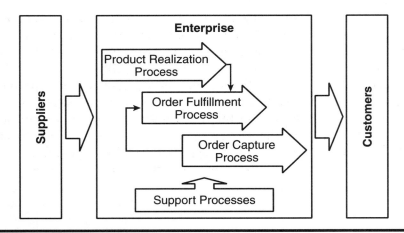

Exhibit 1.1   Typical Enterprise Process Model

The product realization process (PRP) consists of all activities associated with converting market need into a deliverable product or service. This process includes those activities that have traditionally gone under the title of research and development, new product development, product planning, and new product introduction.

The order fulfillment process (OFP) consists of all the activities necessary to convert an order for a product or service into a delivered product or service. This process includes those activities that have traditionally gone under the title of manufacturing, production, logistics, warehouse management, and customer service.

The order capture process (OCP) consists of all the activities necessary to grow, improve, and nurture customer relationships, as well as to convert a customer's need into a purchase order. This process includes activities such as field sales, promotional materials, publishing and distribution, call centers, applications support, and customer training.

Support processes can include a variety of processes and activities, such as field service, administrative support, human resource management, financial management, and legal. The field service process consists of converting an order or obligation for maintenance or repair services into a delivered service. The administrative support process consists of providing the services to the other business processes that are necessary for the successful operation of the other business processes. A human resource management process includes benefits administration, payroll administration, recruiting, and corporate communications. A financial management process includes the internal financial reporting necessary for business decision-making and the external financial reporting necessary to satisfy investors and regulators. A legal support process includes activities necessary to protect intellectual property and to create commercial relationships that conform to all the legal requirements of all the jurisdictions that govern the relationship. To date, there has been a dearth of legal precedent and legislation covering Internet-based activities. It is likely that there will be more legal restraints put on the Internet in the future that will make the legal support process even more important for the e-factory in the future than it is now.

By using this activity-based representation of an enterprise, it is easier to understand and define the key performance indicators that characterize the effectiveness of an enterprise. Examples of key performance indicators include cycle time, throughput rate, activity-based cost, and first-pass yield. Each of these performance indicators can be used for status reporting or real-time feedback control.

Cycle time is an important process parameter because it measures the speed of the process. Cycle time refers to the time it takes to complete one full cycle of a process from beginning to end. An automobile assembly line may be able to build one car in 30 minutes, or a personal computer assembly line may be able to build a personal computer in three minutes. The cycle times for those two processes are 30 minutes and three minutes, respectively.

Throughput rate refers to the total volume of product that can be completed by a business process per unit time. The automobile assembly line may be able to produce 1000 cars a day, and the personal computer assembly line may be able to produce 10,000 machines a day. Throughput depends on the number of production lines operating in parallel and also on whether the process is set up to operate with just-in-time and continuous flow.

Activity-based costing is important because it puts a monetary value on each activity that operates within a process. Thus, not only can the speed of a process be a real-time measure of performance, but also the cost of a unit produced by the process. This also provides a way to convert the time parameter to a cost parameter.

First-pass yield is important because it provides a way to measure the efficiency and quality of the process. First-pass yield is the percentage of product produced without any defects the first time through the production process. Ideally, it should be 100 percent. Many industries approach this level — and many industries do not. Any rework or scrap that results from the product that did not make it through without a defect adds to the final cost of the finished goods.

Extending the Enterprise Model into Supply Chains

Generally, very few manufacturing companies since the Ford Motor Company's River Rouge plant in the 1920s are so vertically integrated that they start with raw materials pulled out of the ground and finish with a product that is delivered to the final user. Most final products are the result of many companies working together through a progression of facilities and conversion steps to produce a final product. This cooperation of companies working through a progression of facilities is called a supply chain. Typically, each individual company or facility within a company engages suppliers to provide it with materials, components, or subassemblies, which it then converts to a finished good, which is then sold to a customer that may or may not be the final user. Within each facility, work is performed within business processes as the model introduced above suggests. By using the same modeling concept, the identity, description, and interaction of business processes can be extended into the supply chain.

The interaction between enterprises actually occurs at many points and between many processes. Illustrated in Exhibit 1.2 is the business process model that encompasses the supply chain.

For example, people within the product realization process of one enterprise communicate with people within the product realization process of a customer enterprise in a collaborative manner for the purpose of capturing the voice of the customer or finding out what their specific needs or desires are about new products. There is also communication on the scheduling and coordination of testing and introduction of new products.

People within the order fulfillment process of a supplier communicate with the order fulfillment process of the customer to ensure that orders are filled correctly and that logistics and delivery coordination is completed. The order fulfillment process of the supplier also communicates with the order fulfillment process of the customer to discuss how to make returns and repairs to products that have been sold and delivered.

The order capture process of a supplier must communicate with a variety of customer processes to ensure that the customer is satisfied with the relationship, the order made, and the settlement completing the transaction.

The support processes of both suppliers and customers must communicate with each other to execute settlement transactions.

This model of the supply chain needs to be converted to a two-dimensional model to fully encompass all the dynamics of a real-world environment. For example in Exhibit 1.3, two dimensions of a supply chain are identified. The horizontal dimension is called the e-supply chain and is discussed in a subsequent chapter section. The vertical dimension is called the e-factory. The e-supply chain refers to the Internet-enabled activities between the enterprises in a supply chain or the external processes. The e-factory refers to the activities between the processes within a manufacturing enterprise or the internal processes that are enabled by or affected by Internet-related technologies.

What Is the e-Supply Chain?

The e-supply chain is the set of external business processes that connect the internal processes of individual enterprises together. These business processes are external to an individual enterprise, but they connect the internal processes of one enterprise to those of another. These external processes embody the relationships and the transactions caused by those relationships that exist between two enterprises. These relationships may have been created by chance or through the efforts of a prolonged negotiation. The transactions may be random or dictated by the specifics of a detailed contract. Whatever the case, these external processes do exist and represent a key aspect of supply chain management in general, and the e-supply chain in particular.

The supply chain process describes the flow of material, information, and transactions between the internal processes of two or more enterprises. It is important to recognize that there is more flowing in a supply chain than just

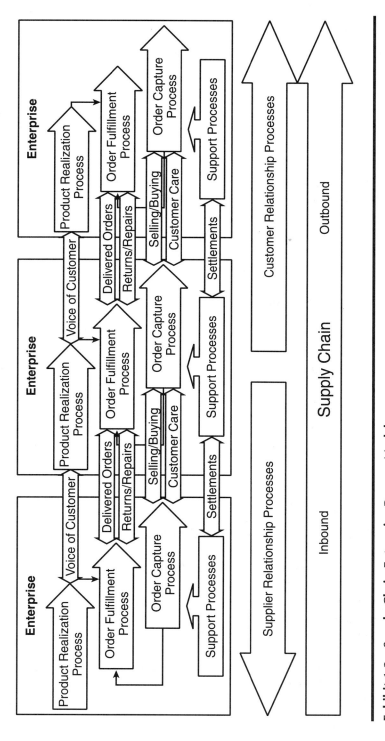

Exhibit 1.2 Supply Chain Enterprise Process Model

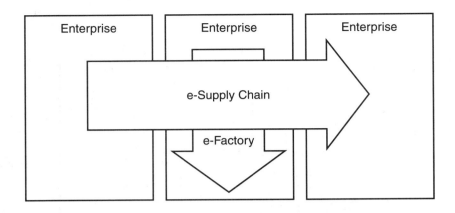

Exhibit 1.3   Two-Dimensional View of e-Business

material. Information is extremely important and includes everything from shipping details and delivery status to detailed quality reports about each item being shipped. The flow of transactions occurs as two enterprises interact with each other and as the information that flows between the two gets processed and actions result. This is why the Internet has become such a potent force in supply chain management. It speeds up the flow of information and transactions so significantly that the nature and characteristics of supply chains are changing dramatically.

The e-supply chain consists of processes and transactions that are controlled and executed electronically. The Internet allows for the rapid flow of massive amounts of information and the rapid execution of transactions. When this information flows in both directions and when the information represents status about existing conditions, then the external and internal business processes can be controlled through feedback. This real-time feedback control capability creates an environment where there can be fewer mistakes in action, fewer miscalculations in status, and faster response in general.

The e-business driven supply chain is more like a network or Web of nodes where each node is performing some type of value-adding activity. Exhibit 1.4 illustrates a supply chain that typifies the semiconductor industry today. Each node in this supply chain network is a company, and in particular, a facility of a company. Information in the supply chain consists of all the interactions between all the nodes in the network. For this supply chain, or in reality a supply web, to operate successfully, all of the transactions, communications, and deliveries between all of the nodes must happen at the proper time and with the proper results.

As an example of this network effect, assume that the end user is buying a cellular phone from a sales company or retail outlet. The end user is buying this phone because of the planning and research of the marketing company that has enlisted the sales company to offer the marketing company's products. The marketing company has outsourced the actual design of the cell phone to a product development company. The product development company, as

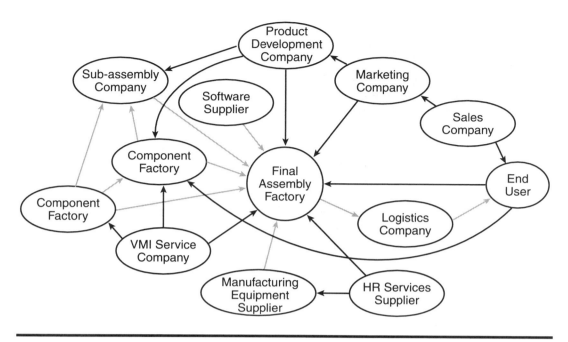

Exhibit 1.4   Supply Chain as a Network of Nodes

part of its efforts, has received information and prototype products from component suppliers upon which it decides whether to include such components in the final design. The product design company must also communicate the design details and the manufacturing process requirements to the manufacturing company with which the marketing company has contracted for final product assembly. The manufacturing company must communicate with and form contracts with the component and service suppliers selected by the design company. The manufacturing company then engages in establishing a schedule with the suppliers so the new product design can be launched and taken to market by the sales company. Of course, with a major new product line, the manufacturing company may need to upgrade its internal software to handle the new volume or the special requirements for information flow dictated by the newly configured supply chain. In this case, the company that supplies the software solution becomes a key element of the supply chain.

In the case of this particular product and industry, each node can be and is a different company. For other industries that are more integrated, each node could be a specific facility within one corporate entity. In both cases, the supply chain operates as a network that must be controlled in real-time via electronic means; thus, the e-supply chain.

Strategy has always been important in the business world, but one of the key strategic issues facing any company in the world of e-business is where will it add value in the supply chain. Using the nodal network model of the supply chain such as in the previous example translates the supply chain strategy question into one of which nodes it will own and which will it connect with via partnerships, subcontracts, or casual transactions. In making these decisions, it must be decided whether value will be added via material conversion, data conversion, or service support in the nodes it owns.

When using a nodal network model for the supply chain, it is clearly seen that factories can be viewed as just a node in the e-supply chain. Inside this factory node are the business processes that must interact with the e-business facilitated external processes. As a result, each process in a factory must be e-business oriented to satisfy the cycle time requirements, to operate within the network of supply chain nodes, and to provide the flexibility and speed required.

In summary, e-business is radically transforming business processes. Industries are disintermediating because of e-business. Enterprises are rethinking where they add value and thinking about their industry as a network or assembly of component processes. As a result, there are significant transformations of entire industries and, more importantly, entire supply chains where there is a regrouping who owns the components and where they fit in the supply chain. e-Business is changing supply chains by changing the structure, the dynamics, and the economics of how large groups of companies work together.

What Is the e-Factory?

The e-factory is the vertical dimension in the two-dimensional e-supply chain. In simple terms, the e-factory is a new, all-encompassing term for all of the electronic control, automation, and intelligent machines that occupy today's factory environment. Electronic control of the factory has been growing in breadth and sophistication for the last two decades. With each new increase in performance and reduction in cost in computer technology, the factory environment comes under greater computer control.

What has happened in the last five years is that there has been a rapid convergence of several trends in the factory. Computer technology has made it possible for every mechanical device in a factory to be intelligent and, therefore, interactive and real-time controllable. Supply chain management science has reached a point where there is philosophy and software that provide for planning policies that deal with short-term and long-term operation of a factory. And the rapid expansion of e-business has created a need for an entirely new set of management requirements for the factory and logistics systems.

A key element in developing a vision for the e-factory is having a model for how the factory can be controlled. A control architecture (see Exhibit 1.5) needs to take into account where control functions are executed and the time requirements for these functions. A control architecture should consist of at least the following levels:

- *Factory level:* refers to the highest level of control for the operations at a facility which could include more than one production line and more than one business process.
- *Line level:* refers to a production line, focused factory through which flows material and information that results in the completion of a

finished product or finished subassembly which is then deliverable to a customer or finished goods inventory.

■ *Cell level:* refers to a portion of a production line within which a variety of activities are performed. A cell can consist of one or more machines that work together or that are organized into a working area where a variety of activities are performed.

■ *Machine level:* refers to a piece of equipment used to perform or assist in the performance of a manufacturing activity on a unit of production.

■ *Unit level:* refers to a unit of production that could be a part, a subassembly, or a final assembly. This is the lowest level of the architecture.

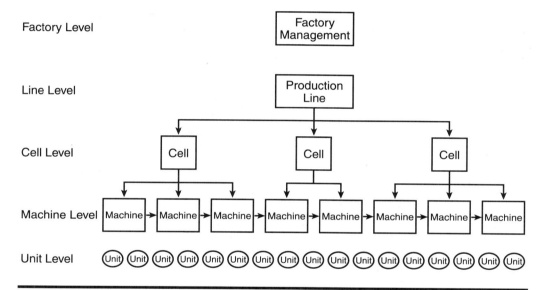

Exhibit 1.5   The e-Factory Control Architecture

The simple description of the relationship of these levels goes as follows. A unit of raw material enters the first step of a production line. At the end of the production, a finished unit of production is completed and shipped to a customer or placed into a finished goods inventory. At each step in the production line, a machine operates on the unit of production. Several machines could be organized into a cell. A cell is necessary when there is a clear physical need to organize all the machines into groups. All the cells together form a production line. Within one factory, there could be more than one production line.

Where Is the Benefit?

There continues to be debate about the benefits and impact of the e-business environment. While there are fantastic claims about global unification, instant business processes, and price reductions due to instant global competition, the reality is that much — if not most — of the benefit from e-business activities will come from more efficient internal and external processes, shorter cycle times in order fulfillment, and small inventory levels.

The benefit of the e-supply chain will be in smaller inventories and faster throughput through the supply chain. This is important because inventory overhangs drive the downside of business cycles deeper and cause the upside of business cycles to experience hot, inflationary increases.

The benefit of the e-factory will be in more productivity, shorter cycle times, and smaller inventory levels. The e-factory will make the nodal network model of a supply chain more practical and more effective. Networks of e-factory nodes will make e-supply chains easier to form and easier to reconfigure. In reality, the e-factory is a key building block in the Internet-driven global economy of the future.

Summary: e-Business Consists of e-Supply Chains and e-Factories

There are two dimensions to the application of e-business technology to the supply chain. The horizontal dimension is herein called the e-supply chain and includes all functions in the inbound-to-outbound business processes that are e-business enabled. The vertical dimension is herein called the e-factory and includes all functions in the enterprise-to-factory-floor that are e-business enabled. There are several points of intersections between the two dimensions that define the nature of the business processes that are external to the enterprise.

The e-factory and e-supply chain are all about breaking down the complexity of an organism of enterprises into controllable elements: supply chains into enterprises, enterprises into nodes, nodes into business processes, business processes into objects and activities.

As discussed, one of the key features of e-business is that business relationships will be executed in real-time. Real-time means that individuals and companies are expecting rapid answers to their inquiries or communications. It is the real-time aspect that will cause the greatest change in all aspects of the economy, especially the manufacturing sector. This real-time response is causing companies to rethink how they manage, control, and organize their internal operations to satisfy externally driven customer processes and externally driven supplier processes. This is causing them to rethink how to organize their logistics systems, warehouses, distribution centers, manufacturing plants, and purchasing operations. This is the essence of the trends that are causing the e-factory to emerge as the most significant trend in manufacturing industries in 30 years. This book is about the e-factory and its interaction with the e-chain, at the activity level, at the node level, at the process level, and at the supply chain level.

Chapter 2

Trends Creating the e-Factory

Purpose of This Chapter

There are a variety of forces that for a variety of reasons are creating changes in the way manufacturing industries operate. Many of these changes have been underway for over a decade and have little to do with Internet technologies. Other changes are only now beginning to take place and have been inspired or stimulated by Internet technologies. Although these changes are being caused by different sources, combined they are defining the e-factory. The purpose of this chapter is to discuss these key trends, describe how they are affecting the evolution of the e-factory, and highlight how they are important to creating a vision of the e-factory.

Collapsing Cycle Times

Using the enterprise and supply chain business process model introduced in Chapter 1, the key parameters that describe the properties of each business process include cycle time (the time for one unit to go through the process), throughput (the number of units produced by the process per time unit), first-pass yield (percentage of units successfully produced by the process on the first pass before rework or scrap), and by activity-based cost (the cost to produce one unit). Exhibit 2.1 is an illustration of the different types of cycle times by business process.

These parameters are, of course, related because throughput is inversely related to cycle time and cost is inversely proportional to yield and throughput. If a business process is improving, the key performance measurement parameters should indicate that cost is decreasing, yield is increasing, throughput is increasing, and cycle time is decreasing. As will be discussed in a later chapter,

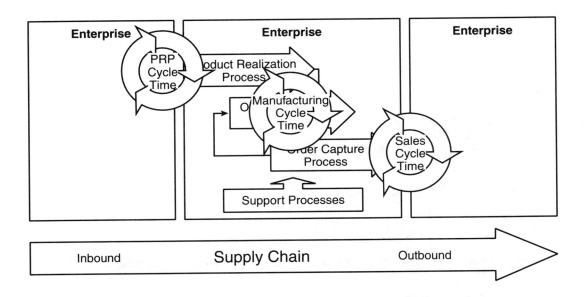

Exhibit 2.1   Supply Chain Cycle Times

all of the things that can be done to improve throughput, cost, and yield also improve cycle time, and vice versa. All of the effort invested by companies in every manufacturing industry to improve business processes has also resulted in a significant improvement in cycle times.

The cycle times of the key supply chain business processes are all trending downward. For the last two decades, significant effort has been invested in reducing manufacturing cycle time by eliminating non-value-adding activities, combining or consolidating activities, performing activities in parallel rather than in series, controlling critical processes to within smaller tolerance ranges, and using technology to accelerate critical-path activities. In virtually every manufacturing industry, there have been significant reductions in manufacturing cycle times, ranging anywhere from 50 percent to 90 percent or more. In the 1980s, many manufacturing industries had manufacturing cycle times characterized by months or weeks. Today, these same industries have manufacturing cycle times of days or hours. A chart of the trends in process cycle times is given in Exhibit 2.2.

Product development cycle times have also kept pace with the reductions in manufacturing cycle times. In many industries, product development cycle times used to be measured in years. Today, they are measured in months or even in weeks. These product realization process (PRP) cycle time reductions have come about due to the greater use of computer systems that accelerate product and process design, modeling, simulation, product data management, and cost modeling. In addition to technology, the processes that companies use for PRP have improved with the use of cross-functional teams working on several activities in parallel, with decision support tools such as quality function deployment to accelerate the definition of product specifications and identification of voice of the customer, and with product family strategies that build on and reuse design data.

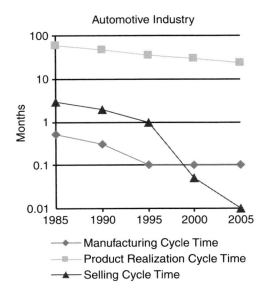

Exhibit 2.2   Trends in Key Process Cycle Times

While people have been working on manufacturing and product development cycle times for years, very little was being done to improve selling cycle times until just recently. However, reductions in selling cycle times could be the greatest factor in creating the e-factory. Over the last five years with the advent of e-business, sales force automation, customer relationship management software, and on-line Web site-based shopping malls available from virtually every retail business and from many manufacturers that have never sold directly to consumers, sales cycle times have dropped drastically. The sales cycle time is no longer measured in months or weeks but now in terms of mouse-clicks. With the cycle time measured in seconds or minutes, manufacturing cycle times measured in hours or days are now no longer the shortest cycle times in the supply chain but are now becoming the longest. This has a tremendous impact on the strategy and planning for manufacturing that leads to the need for and the concepts associated with the e-factory.

As the sales process cycle time shrinks, there is a great need for manufacturing cycle times to shrink further so that the order fulfillment process is able to respond to shifting market trends and to meet customer expectations for fast order fulfillment. The collapse of selling cycle times is creating major changes in what is expected from the manufacturing cycle time.

When manufacturing cycle times are shorter than selling cycle times, factories can build to order, reschedule production very frequently, and operate with very small inventories, which means a high inventory turnover rate. When manufacturing cycle times are longer than selling cycle times, factories must build to forecast, increase inventory stocking levels, and schedule production over relatively long periods of time. This decreases flexibility and increases the chances of overstocking slow-moving items and understocking fast-moving items.

Innovation in Supply Chain Management

In addition to the dramatic reduction in sales cycle time stimulated by the Internet, there have been a variety of innovations in the general area of supply chain management that have evolved over the last few years that by themselves contribute to the need for e-factory capabilities. These innovations include extension of enterprise process improvement to the supply chain, fission of the supply chain, fusion of the supply chain, and collaboration within the supply chain.

Enterprise Extension

Enterprise extension refers to the application of the business process best practices and design concepts that work so well internally to the supply chain operations that are external to the enterprise. All of these extension ideas have evolved from the same goals that manufacturing companies have shared for many years: (1) minimization of inventory, (2) reduction of cycle times, and (3) flexibility in responding to market needs. By setting these same goals for operations in the supply chain and using many of the same techniques, significant overall improvements can be achieved in the supply chain.

One of the key drivers in the improvement of business processes is the minimization of inventory. Closely tied to this goal is reduction in cycle time. It may be inherently obvious that short cycle times are good and high inventory is bad, but it is worth examining the basic thinking underlying these statements. Inventory represents the investment of cash in finished goods or work in process. The larger the investment, the higher the risk that the inventory could become obsolete or of reduced value. Cycle time for the production process and cycle time for the product life both play a role in determining the degree of risk. Exhibit 2.3 illustrates the relationships of cycle times to risk for inventory investment. The financial risk, the risk that some portion of inventory will become obsolete, is inversely proportional to the number of inventory turns. The number of inventory turns is inversely proportional to the inventory level. Thus, at the highest levels of inventory (i.e., the lowest number of inventory turns), the risk is quite high that a significant percentage of the investment in inventory will be lost.

Applying this thinking to the supply chain, an even greater financial leverage can be observed. The diagram in Exhibit 2.4 illustrates the use of a simple supply chain model for inventory lead times and manufacturing cycle times that could happen in the build-up of inventory throughout the supply chain. In one set of data, the plot indicates that the inventory in each link backward into the supply chain builds up exponentially because of the need to forecast the demand at each link with the appropriate safety factor. In the other set of data, there is communication throughout the supply about the demand at the final link in the supply chain. This demand communication allows each link in the supply chain to know more precisely what the order level will actually be. This mitigates the need for excess inventory and

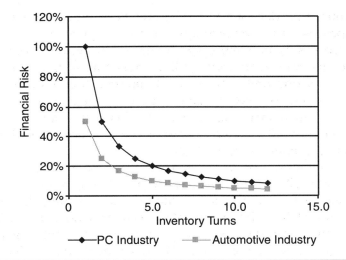

Exhibit 2.3   Relationship of Inventory to Financial Risk

excessive minimum inventory levels. Thus, the financial risk inherent in the supply chain is reduced. This reduction in financial risk in supply chain inventory has been a source of stability in the U.S. economy over the last decade by minimizing the inventory overhangs that were present in inflationary-recessionary economic cycles.

Demand communication evolves from the concepts associated with just-in-time and pull manufacturing philosophies. In the simplest terms, the concept of pull manufacturing is that when a unit is sold, a unit is then made to

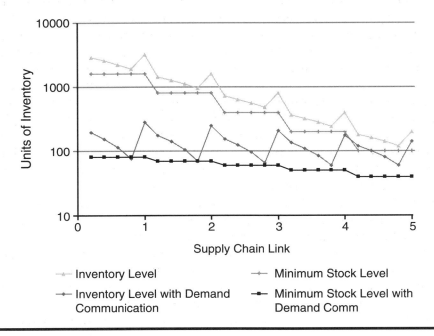

Exhibit 2.4   Supply Chain Inventory Model

replace it. Thus, an order for a unit of production "pulls" product through the production process. The opposite approach is to "push" product through the production process by building up an inventory based on a forecast of future sales. In the extreme ideal case, pull manufacturing results in an average inventory level of one unit. Obviously, communication about what is selling and what is needed to meet demand (i.e., demand communication) is very important in pull manufacturing because the fact that a unit was sold or used must be communicated backward throughout the order fulfillment process and ultimately the entire supply chain.

Demand communication rarely makes much of a difference in process performance by itself without having order fulfillment processes structured to take advantage of the information communicated. During the last 20 years, the companies that have benefited most from demand communication principles did so by physically restructuring their order fulfillment processes to operate in a just-in-time, continuous flow manner. The production process was physically laid out to have a physical flow of material that had a minimal amount of wasted motion and that was visible to all the people in the flow. Visibility of material was important because it allowed demand communication to occur visually. Typically, production processes were organized into cells. Each cell had a lot size that was calculated to be a function of the speed and capacity of the cell but was as small as possible, which was ideally one unit. The output of the cell, a very small lot size, was placed in a very visible spot next to the cell. Another lot was not processed by the cell until the finished lot was pulled away by the next cell in the production process. The communication of the demand for product being placed on the total production process became visible by the very absence or presence of finished goods in the output space next to each cell. For small production operations with only a few cells, this physical visibility is enough. As the number of cells increases or the production process begins to cover more physical space, communication aids such as kanban cards or other communication forms can be used to signal that a cell needs to pull a material lot into its operation. To extend the benefits of pull manufacturing into the supply chain, demand communication must occur over larger distances. This requires a coordination of physical layout and operations management, which calls for more electronic communication. With e-factory technology, virtual kanbans can be created in the computer systems controlling the processes and then communicated electronically to all relevant members of the supply chain.

Demand communication has been only one type of innovation in supply chain management. Reducing the cycle times of the key business processes is an equally important factor in improving the performance of the extended enterprise. If process cycle times are long, then lot sizes must be large, and much of the benefit of demand communication is diminished. If process cycle times are short, lot sizes can be small and ideally approach a lot size of one unit. With short cycle times, demand communication becomes not only beneficial, but also essential to the successful operation of the supply chain.

To achieve the shortest cycle times in business processes that are internal to enterprises, there are some general best practices for structuring a process, including:

- elimination of non-value adding activities
- performing non-critical path activities in parallel
- accelerating key activities with technology
- controlling activities to minimize variation
- reduction of setup times

When these cycle time reduction actions are applied to business processes that extend into the supply chain, major savings in not only inventory but also cost are achieved. The entire supply chain typically becomes much more flexible as well. In many cases, it is the extension of these cycle time reduction capabilities that creates the opportunity for innovative approaches to planning, managing, and designing a supply chain. It also creates new phenomena in the supply chain such as fission, fusion, and collaboration.

Another example of extended enterprise process thinking into the supply chain is in the management of warehouses and distribution centers. Whereas warehouses and distribution centers were once viewed as just storage areas in an overall logistics network, they now have to be viewed as critical-path, order fulfillment centers. e-Commerce is now driving final customer order fulfillment back into the supply chain into distribution centers or factory warehouses. Final order fulfillment that used to take place at a retail store is now being done farther back in the supply chain. On-line orders need to be picked, packed, and shipped in real-time, which means that large warehouses now must become fast, small-order fulfillment centers.

In addition to order fulfillment, warehouses now must be fast and efficient returns processing centers. As the number of on-line orders being filled increases, the number of orders being returned to fulfillment centers increases. While a true believer in the principles of total quality would say that if the customer could be satisfied every time, there would be no need for returns, the reality is that e-business technology has not reached the point where customers can be totally assured that what they are ordering is exactly what they want.

The new form of warehouse management is an extension of enterprise concepts because warehouse order fulfillment processes now must have the same characteristics as the production order fulfillment processes. Warehouse management now includes the need for real-time visibility into exact inventory levels for large numbers of stock-keeping units (SKUs), the need for replenishment practices based on much shorter cycle and lead times, the need for order status tracking, and the need for tracking smaller lot sizes.

Traditionally, distribution facilities have managed large incoming receipts of product by the case, carton, pallet, crate, truckload, or container. These deliveries are then put away, stored, and later picked and packed into outgoing multi-line orders for shipment to retailers, suppliers, or third-party warehouses.

Today, warehouses are responsible for handling many more receipts of smaller quantities of customized or niche products and moving high volumes of single-item, on-line orders, often shipped via overnight freight or package delivery service direct to the consumer.

Supply chain planning is an extension of the planning that was necessary for the production processes contained within one facility or enterprise. The mathematical techniques used to plan and schedule for multiple workcenters with finite capacity have become the same tools used for planning and scheduling multiple plants and warehouses in a supply chain. The concept of multi-plant scheduling within a supply chain became more practical when software tools providing this capability were introduced to the marketplace in the late 1990s. These tools applied advanced mathematical techniques to provide solutions for the scheduling of multiple plant situations.

A final example of extending best practices beyond the bounds of the enterprise is in product and production process design. For the last ten years, there has been case after case of companies announcing that they had discovered the benefits of DFX. DFX is a term used to refer to a genre of computer-aided design tools that specialize in a variety of capabilities such as design-for-manufacturability, assembly, test, simulation, engineering, etc. By taking into consideration the manufacturability, service, and fit-for-use issues early on in the product development cycle, many of the quality and cost problems that often plague new product introduction to manufacturing can be eliminated. The obvious next step in that type of thinking is to extend it to the concept of the supply chain. By taking into consideration the supply chain management issues early on in the product development cycle, the problems that typically plague the introduction of a new product to the market can be eliminated. The extension of DFX into the supply chain requires new levels of collaboration in product realization that go beyond even the collaboration that should be taking place within an enterprise. The innovations occurring in this area are very promising and will likely be a key part of the future evolution of the e-factory.

Fission

Extended enterprise thinking immediately leads to the concept of fission and fusion in the supply chain. *Fission* refers to the dissolution of an existing entity into multiple smaller entities. As manufacturing and supply chain strategists analyze how to improve the operation of a set of business processes that operate outside the boundaries of the traditional enterprise, many new approaches to executing the same set of processes emerge. Many of these new approaches include transferring key activities or business processes to new entities outside the boundaries of the traditional entities or enterprises.

One form this transference often takes is outsourcing. There has been a major trend toward outsourcing in several industries over the last ten years. Industries such as the semiconductor industry and the Internet equipment industry have been extremely innovative in implementing outsourcing arrangements with contract manufacturers, engineering design companies, marketing companies, and tooling companies. In fact, the trend in these industries has been rapidly evolving to virtual supply chains where what would have been considered an order fulfillment process contained within one company in the past has been decomposed into multiple companies in different parts of the world.

The types of functions and activities being outsourced today include fabrication and assembly, maintenance and maintenance management, inventory management, design, marketing, and even operations management.

Contract manufacturing has been in existence for decades. However, the concept of outsourced manufacturing has evolved to a point of sophistication that involves reengineering of internal and external business processes. For example, Cisco Systems contracts with a manufacturing supplier (e.g., Solectron) to produce a certain number of units while maintaining control over scheduling, component suppliers, and logistics.

A newly emerging outsourcing service is maintenance management. Companies that produce the equipment or materials used in manufacturing by certain industries are now offering the service of being totally responsible for scheduling, stocking, and servicing the repair and maintenance of the equipment they make, as well as the other equipment operating in the customer's production process.

This concept of maintenance management is happening in parallel with the trend in vendor-managed inventory. In vendor-managed inventory, a supplier of a material or component will monitor the stocking levels for its parts at its customer's sites and will restock inventory when it drops below a minimum level. This solves the demand communication problem for the customer site because the supplier of the material being stocked is looking forward into the supply chain to see what demand is doing to stocking levels. This also eliminates work at the customer's site because no labor or effort is needed to manage the supplier's inventory. New innovations are being brought to market for the e-factory by suppliers who are installing Web-enabled monitoring devices on their equipment at the customer site, which alerts the supplier when inventory has dropped below the prescribed level. This demand signal then causes a replenishment action for the inventory. In the past, the supplier would have to send a person on a regular basis to manually inspect the inventory levels at each of their customer sites. Now it can be done in real-time.

Another form of fission occurs when there is an entire transformation of all the companies involved in a supply chain to the point where they become specialized nodes in a supply web. This nodal transformation occurs because of the new opportunity for companies to rethink where they add value in a supply chain and how they can focus their efforts as an e-factory node or as some other type of specialized enterprise node.

Fusion

The other direction that extended enterprise thinking leads to is fusion in the supply chain. As e-commerce and e-factory technology makes it possible for a manufacturing company to expand its capability for providing customer satisfaction forward and backward into the supply chain, the opportunities for fusing (or integrating) these capabilities into its normal operations increase. While this appears to be contrary to the other direction that extended enterprise thinking is taking toward fission, it is in many ways the trend that is accelerating the advent of the e-factory the fastest. Examples of trends in supply chain fusion include logistics, planning, and product development.

Logistics fusion refers to having one supplier responsible for all logistics issues, such as all forms of transportation, managing inventory, and tracking shipments. Fusion in logistics processes is leading to companies that are offering these types of integrated services and are thus becoming logistics nodes in a supply web.

Planning fusion refers to having all companies that work together in a supply chain as customer and supplier jointly develop a planning mechanism for the flow of goods and services between them. Fusion in planning processes is leading to companies that are offering these types of integrated services and are thus becoming planning nodes in a supply web.

Product development fusion refers to the joint development of a new product. One of the best practices in new product development is the use of cross-functional teams. This concept of cross-functional teamwork is being expanded to include people from suppliers or even customers. Fusion in product realization processes is leading to companies that are offering these types of integrated services and are thus becoming product realization nodes in a supply web.

Collaboration

The contradictory trends of fission and fusion do create a third trend that is actually more dramatic and transformational in nature and may not be sustainable. This third trend is collaboration.

Collaboration refers to the cooperation of multiple companies in a common endeavor. While supply chain cooperation has been growing as a result of virtually all the trends mentioned above, changing cooperation to outright collaboration is a quantum step in change. Collaboration implies that companies that have a vested interest in not sharing data, tactics, or strategy, actually do share data, tactics, or strategy.

Trends in greater collaboration are occurring in product development, demand management, purchasing, and selling. It is the consideration of the use of collaboration that has been the hot recent trend, not necessarily the actual implementation of collaboration. For collaboration to become a major behavioral trend, the legal and emotional barriers related to conflicting vested interests must be overcome. The legal issues include anti-trust legislation, intellectual property ownership, and new forms of redress if there are disagreements after collaborative relationships are created.

Product development collaboration refers to the sharing of needs, requirements, process capabilities, previous history, and future plans. Creating cross-functional product development teams that include customer and suppliers alike has been a best practice for years. However, a more intimate form of cooperation might be a major challenge. Clearly, computer tools such as workgroup technology, product data management, generative design technology, and product configurators that assist in the creation of new product solutions are improving in their ease of use and comprehensiveness. Along with these improvements come a greater array of teamwork features.

Demand management collaboration refers to the sharing of information about supply chain demand, both actual and forecasted, at multiple points in the supply chain. As discussed, the communication of demand trends has been a major breakthrough in the management of inventory and response time in the supply chain. To take that communication a few steps further and share the strategy and tactics related to demand creation falls into this new domain of demand collaboration. It is the obvious next step to supply chain planning and demand communication. However, the legal and proprietary conflict barriers must be overcome before further evidence of demand collaboration will become visible. These issues are now being addressed by the newly formed e-market consortia formed during the year 2000 as several large companies decided to put all of their buying and selling on the Web.

Purchasing collaboration refers to the sharing of purchasing efforts to achieve better prices, better terms and conditions, and better service. There has been a flurry of activity by several companies in different industries to pursue new approaches to managing their purchasing activities using Internet-based technologies. The consortia that have been announced in some ways are examples of e-commerce collaboration and in other ways more like an electronic version of the industry cooperatives that sprang up during the last century among small businesses that had a common interest, such as small farmers and small investors.

Selling collaboration refers to the sharing of selling efforts to achieve better prices, more market share, and greater penetration of new markets. The selling channels have seen the greatest amount of turmoil in terms of transforming how companies go to market. There has been a flurry of activity here as well by several companies in different industries pursuing new approaches to pooling their selling activities using Internet-based technologies. The consortia that have been announced in some ways are collaboration and in other ways are more like creating or aggregating electronic catalogs of products that can be quickly searched by Web-based technologies.

As the computer tools that enable cross-functional teamwork and external enterprise cooperation in all of these areas continue to evolve, they are touting collaboration as a major breakthrough. Whatever final form collaboration takes, it will be a major influence on the e-factory because it will require that the e-factory be able to support the collaborative links both electronically and physically.

Ubiquitous Computer Technology

Of course, a discussion of trends affecting the e-factory cannot be complete without a treatment of technology trends. Virtually all the technology trends that have created the e-business environment contribute to the e-factory. Computer and communication technology is now ubiquitous. Personal computers now appear on every desk, at every worksite, and within every expensive machine. Computers are no longer shackled by the wires of power or communication. Handheld devices, whether they are cellular phones, personal digital assistants, or personal computers, are now becoming the e-commerce platform in Europe and Asia. Almost any combination of mobility, computing power, communication volume, or media format is now commercially available.

In addition, the advent of the Internet and the World Wide Web has created the reality of virtual networks with multiple entry points. No longer does an e-factory have to be constrained to a wiring topology or narrow bandwidth limitations. Factory control architectures do not have to be built around physical or computational constraints.

This ubiquitous technology has created an environment in which virtually every machine, every material handling device, even the unit under production, can have intelligence that provides for real-time information, real-time control, and interactive on-line communications. This environment eliminates whatever technology limitations there may have been in the past and has created an entirely new realm of opportunity to design and configure the e-factory.

Mechatronics: Intelligent Machinery

Given that technology is now ubiquitous, an additional driving force behind the evolution of the e-factory is the explosion of electronically controlled and actuated mechanical machines called "mechatronics." The concept of mechatronics is important because it incorporates a variety of capabilities that are critical to the e-factory. These capabilities include the integration of mechanical motion systems and actuation technology. Actuation technology includes electrical, magnetic, and fluidic powered actuation. These capabilities include the integration of feedback control into actuated motion systems. Feedback control is based on sensors, computers, and control software. These capabilities include flexible automation such as robotics, advanced sensors, high-precision material handling systems, and flexible automated cells.

Mechanical systems have been a key part of the manufacturing environment since the advent of the industrial revolution. The extent to which machine power is used to replace or multiply human power has been the ultimate measure of an economic system's performance for the last century. Every time there has been a major innovation in the use of machines — whether mechanical or electromechanical or electronic — there has been a significant increase in economic productivity and prosperity. The emergence of the e-factory is due in large part to the major mechatronic innovations that are creating a new burst of productivity and prosperity.

During the last few years, there has been a major renaissance in the design and application of motion systems. The oldest forms of motion systems are those that move material through a production line, handle material at critical points by moving it from one workcenter to another, present material to a workcenter, or handle material while it is in a workcenter. Modern applications of motion systems are not only for moving material but are also for moving machines and people within workcenters. The e-factory must be not only efficient but also flexible. This means that the entire configuration of all the resources in a factory — people, machines, and data — must be changeable in less than a few multiples of the cycle time of the supply chain. As it turns out, the convergence of the evolution of computer systems, industrial electronics, and mechanical motion systems results in practical mechatronic systems that create this level of flexibility.

A mechatronic system consists of motion-facilitating mechanical devices, electronic controls for the devices, and actuation for the devices. The motion systems typically consist of linear and radial, bearing-based assemblies that are actuated by electrical motors, hydraulic pistons, or compressed-air pistons. Action is controlled by some type of electronics, such as programmable logic controllers, custom electronic systems, or personal computers. Originally, mechatronic systems were clever assemblies of the individual components. By using DFX tools, they are now becoming integrated modules that have maximum capability with minimum cost. Modular integration of mechatronic technology is making it possible to use the best configuration of software, electronics, mechanics, and packaging to provide the best solution.

The motion components are either linear or radial motion. Radial motion components typically consist of bearings that are designed to facilitate rotation of a mass around a central shaft or point mass. Linear motion components typically consist of bearings that are designed to facilitate back-and-forth motion over a shaft or rail. Multi-axis linear motion systems are constructed by combining multiple single-axis linear motion systems. Multiple degree-of-freedom systems can be constructed by combining multiple single-axis linear motion systems with radial motion systems.

Actuation of motion systems typically consists of some type of electrical energy source that is converted to mechanical energy that then becomes mechanical motion. Energy conversion approaches include electric motors, hydraulic pistons, or compressed-air pistons. Electric motors can generate either rotating mechanical motion that can be turned into linear motion via a mechanical screw, or electric motors can be designed to produce linear mechanical motion directly. There has been a steady flow of innovations in the integration of electronic controls into electric motors that are creating new opportunities for more flexible and efficient actuation, which increases the opportunities to use automation in broader e-factory applications.

The development and application of industrialized electronic control computers to control the actuation of the motion components has been a major advancement in the productivity and flexibility of automation systems. Using feedback control systems to control motion systems provides the maximum in precision, repeatability, and flexibility. For many years, the control of mechanical systems was open loop; that is, the mechanical parts were designed to be of such precision that their movement was repeatable to within certain tolerances for many hours of operation without any corroborating measurement with external sensors. Although the best motion systems are designed using frictionless bearings, mechanical wear does occur. After thousands of hours of operation, mechanical parts would have to be replaced or refurbished in order to re-attain the level of precision as originally designed. With feedback control systems, the actual position or movement of motion systems can be measured using some type of absolute or relative reference. The correct final position can then be calculated and control instructions communicated to the actuation system to move the system to the final precise position. Virtually all navigation, guidance, and control systems for high-value or high-cost systems use feedback control. It is only recently that the use of feedback control systems for common applications of motion control systems has become cost-effective.

Not only does electronic control provide for greater precision and repeatability, but it also provides for greater flexibility. Although great care must be taken in designing the mechanical system of motion components, the extra dimension of flexibility is provided by the use of programmable control systems. In the early years of robotic systems development, there were many innovations in programming languages for robotic control systems that focused on providing the appropriate macro commands that would fit with the language of the multi-axis control problems associated with flexible automation systems and robotic systems. Over time, these program languages have become much more user-friendly and are more flexible themselves.

The entire concept of electronic control has evolved to the point where control-by-wire concepts are now being applied to consumer applications. Control-by-wire is defined as a control system that takes control commands from one physical location and communicates them electronically to an actuation system in another location for final execution of the command. Commercial and military aircraft have been control-by-wire for many years after replacing pure mechanical linkages between controllers and actuators. When the pilot manually turns or moves the steering yoke in the cockpit of an airplane, electrical signals are created by sensors attached to the steering column and transmitted to the many electrical and hydraulic actuators throughout the plane which then move the appropriate wing surfaces or other devices.

Replacing mechanical linkages with electronic control-by-wire systems is beginning to emerge in industries other than aerospace. The automobile industry is a prime candidate for such a transformation. When a driver turns the steering wheel in an automobile today, there is a hydraulically assisted mechanical linkage that turns the front wheels. Likewise, when the brake and accelerator pedals are pressed, there are mechanical linkages that communicate and convert the pedal motion into actuation at the brake or fuel control devices. Because of the potential cost savings, these linkages will likely become electronic linkages in the future.

Control-by-wire applications offer performance improvements as well as cost advantages. For example, the use of control-by-wire electric power steering offers improved fuel mileage because the system is lighter than the hydraulic system it replaces, and it uses less of the energy provided by the engine. In a hydraulic power steering system, the engine has to drive a pump that runs constantly to keep the appropriate pressure on the hydraulic fluid in the hydraulic system; whereas in an electric power steering system, which is control-by-wire, the only time energy is used is when the steering wheel is actually being turned.

As the application of electric actuation has grown over the years, innovative motor designs have evolved to allow for more practical and effective applications. Brushless motor designs have emerged as a popular technology because they offer longer lifetimes since they do not have brushes that are constantly being worn down due to contact friction. While brushless motor design concepts were invented decades ago, their practical application has been limited due to the need to use computer control to perform the necessary electrical commutation and operate the motor successfully. Low-cost microelectronic technology now makes this motor design much more attractive.

A second major innovation in motor design has been the linear motor. A linear motor is nothing more than a brushless motor that has the stator laid out on a flat plane rather than in a cylindrical shape concentric with the rotor. This architecture allows for the magnets in a linear motor to be collinear with the path of the linear motion and of unlimited length. Linear motor systems can replace conventional electric motor ball screw designs for certain applications. The benefits of linear motors are that they offer more acceleration, increased smoothness, and less noise than conventional electric motor ball screw designs. However, there is a trade-off of cost versus speed for applications that do not require low noise or extreme smoothness. There are also limitations for the application of linear motors in terms of geometry and fail-safe mechanisms.

Control Systems Innovation: Sensing and Responding to Reality

Control systems technology has been evolving over the last 20 years to the point where there are many decisions to make when it comes to deciding on a control architecture for the e-factory. There has been an emergence of new control technology at both the low end of the control hierarchy as well as at the high end, the two of which are now beginning to merge. It is the merger of these technologies that is creating the capabilities necessary to support the needs of the e-factory.

From the bottom of the hierarchy, there have been a variety of control innovations that are creating new capabilities. The programmable logic controller (PLC) has been the mainstay of factory floor control systems for decades. These are the devices that were originally designed to provide the conversion of an electrical control signal into a mechanical or physical action. PLCs are used to do everything: turning on and off valves, actuators, pumps, doors, alarms, and other machines. In their first generation, PLCs were electrome-chanical switches; today, they are personal computers with custom semicon-ductors and circuit designs. PLCs can be found sprinkled throughout a manufacturing plant and usually networked together.

Machine controllers are also at the bottom of the hierarchy. These are the control computers that the maker of a piece of manufacturing equipment would incorporate into the machine design. These used to be custom electronic devices with capabilities limited to sensing and controlling basic machine functions. Today, machine controllers are usually enhanced personal comput-ers that have specialized control software that communicates with electronic sensors and computer networks.

Other intelligent devices are also at the bottom of the hierarchy; for example, optical bar code scanners, radio frequency transmitters, and other forms of intelligent tags. The same technology that is making credit cards smarter and tollbooths drive-through is also making it possible to put inert or active electronic tags on material or material containers. This "intelligent" material can then communicate with the control system as it moves through the business process, providing the ultimate in real-time information on location and status.

At the top of the control hierarchy is the emergence of software designed to run the entire plant or production line. This level of control includes the control of all the resources employed by the plant, monitoring the status of the key operations, collecting data about costs and environmental factors, and making decisions about scheduling and allocating resources in real-time.

Cell controller software was the first technology to be applied at the higher end of the hierarchy. This software was developed to monitor and control only a few pieces of equipment operating in a work cell or workcenter. The first cell controller software packages had unique and proprietary programming languages and were usually designed for expensive computer hardware.

Manufacturing execution software (MES) was the next level of software to be developed. This software was designed to manage more than one cell or an entire production line. The management functions of MES included moving material from point to point in a plant, sending recipes or machine instructions to each piece of manufacturing equipment, and tracking the cost and status of the material being processed throughout a plant.

Eventually, the enterprise software companies brought high-level manufac-turing control software to the market as part of their enterprise solution. This software was designed to provide an integrated — although proprietary — environment that operated at both very high control levels (work order processing, inventory management, labor management, and plant scheduling) as well as at the lower levels. A key part of their product strategy was to minimize the difficulty of integrating control technology at the different levels of the control hierarchy.

Integration of the control technologies at the different levels of the hierarchy must be done. The enterprise software companies did have the right design goals by providing a software product family that was integrated because of a generic control architecture. However, the real world of the e-factory consists of many machines from many suppliers that have many different strategies for control and data collection. Integration of technologies in the factory environment has been a major challenge for years. The requirements for control in the e-factory are increasing the magnitude of the challenge.

However, the advent of the Internet and other relevant technologies is making the integration challenge more amenable to solution. The control technologies at all levels of the hierarchy are being made Web-enabled. This offers many new options for how to configure the hierarchy itself, how to design the communication network between the control and intelligent devices in the factory, and how to integrate the devices with the internal and external business processes. New Internet-driven concepts about database integration and data management are all contributing to making the e-factory easier to implement.

Being able to define the integration path is one of the key requirements of any roadmapping technique for the e-factory. A well-designed, integrated control hierarchical control system will collect data on the factory floor or at the point of activity. It will convert the data to management information in real-time. This real-time information can then be used to make real-time decisions about the employment and configuration of the resources contained in the e-factory.

There are many benefits to using control systems in the e-factory. Feedback control provides for real-time changes and precision (smaller lot sizes), rapid reconfiguration and rescheduling, capacity modeling and planning, status monitoring, maintenance management, statistical process control, and cost data collection.

Control systems also help solve the e-business challenge by being applicable to business processes other than those found on the factory floor. For a business process to be controllable, control systems must be well-defined, the key control parameters must be observable and measurable, and mechanisms must be in place that can make changes to the business process when a control action has to be taken. As business processes become more external and sharable with customers and suppliers, control issues become a matter of concern for all parties in the supply chain.

Cellular Manufacturing

If there were no booming e-business economy or if there were no Internet, the trends in productivity improvement and the opportunity to use state-of-the-art computer technology to accelerate those trends would still create a need for the e-factory and all the techniques and technologies that go with it. Much of the process structure and technology promise that is now readily available for the e-factory was inspired during the 1980s as the first wave of computer-integrated manufacturing (CIM) technology was brought to market.

In many ways, the culmination of all the advanced concepts that were hypothesized or developed at that time are only now becoming commercially viable.

The cost and capability of computer and communication technology have advanced to the point where the early dreams of the e-factory are now practical. There is also an entire body of knowledge of how to effectively maximize productivity using human, data, and machine resources. Without this valuable knowledge, the pace of change in the introduction of new technology would be much slower. This knowledge is most frequently applied in the form of cellular manufacturing.

The definition of cellular manufacturing has many variations. This term is being used herein to refer to a tight organization of manufacturing resources that works together as a team to produce a family of products. A cell should have a minimum of non-value-adding activities, should perform noncritical-path activities in parallel, should accelerate key activities with technology, should control activities to minimize variation, and should have very short setup times. A cell could be a node in a supply chain or a node in an internal order fulfillment process.

The concept of the cell is important because of what it implies. A cell is a production capacity that has been stripped down to its most essential elements and organized for extreme efficiency and flexibility. A production line can consist of more than one cell. A supply chain can consist of multiple cells from multiple production lines. The cell becomes a building block at the molecular level of a supply chain. Technology is an important enabler for a cell. Process design is the backbone for establishing the cell architecture that can be enabled by technology.

Using the cell design concept is an important trend in the e-factory because it provides several benefits. It provides for flexibility in operation because it has been designed to be flexible across a family of products. Flexibility is achieved with short setup times. It provides for a minimum of risk because it is designed to operate with a pull manufacturing approach that is based on small inventory levels. It has been designed to have a short cycle time because of the continuous flow layout and the technology applied. It provides for supply chain flexibility because it can be moved to a new location or become a part of a virtual reconfiguration of a virtual supply chain. The concept of cellular manufacturing is at the heart of the e-supply chain and the e-factory.

Tribal Globalism

While all the trends described above point to greater globalism in competition, market opportunities, and cost management, they also create the opportunity for being able to focus on the tribal nature of customers and markets. The term "tribal" is being used here to refer to market segmentation by demographic or cultural parameters. One of the most frequently used terms to describe what is happening in supply chain management is the "customer-centric supply chain," which basically means that a supply chain should be able to uniquely satisfy each individual customer. However, individual customers tend to congregate into groups that begin to behave like virtual tribes. A major reason that customer centricity is possible is that while the Internet makes it easier to reach global customers and suppliers, it is also making it easier to focus on individuals and affinity groups. For example, the ultimate example of a self-organizing tribal activity is the chat room, which was an Internet innovation.

From a global perspective, the trends in supply chain theory and e-factory technology are making it easier to address global issues and opportunities. Global supply chain cost models can be developed to help define where the nodes and segments of a supply chain should be located. How to build supply chains that take advantage of global resources but which serve local or virtual tribal market needs can also be determined with the new knowledge and tools becoming available. As supply chain practice continues to evolve to address global economic factors while also serving local or virtual tribal market needs, the role of the e-factory as a node in a configurable supply chain will continue to grow in importance.

Summary

The trends creating the e-factory are coming from a variety of sources. The trends eminating from Internet-related technologes are bringing e-business cycle times to the factory floor. These trends are also creating the need for the manufacturing enterprise to be in direct connection with the end users of its products, as well as with any channels that may have connected them in the past. However, other trends being driven by technology advances in a variety of automation fields are also contributing to the great potential of the e-factory. Internet-based trends are creating the need for very short cycle times and major transformations of the external business processes. Automation technology-based trends are creating the opportunity to dramatically reduce the cycle times of internal processes and to make the manufacturing environment as flexible and as productive as the e-business environment requires. Combined, these trends are creating a unique opportunity for the e-factory to emerge from the practices and philosophies that have governed general manufacturing industries for decades.

Chapter 3

The Role of Technology in the e-Factory

Purpose of This Chapter

The purpose of this chapter is to describe several of the more important roles that technology is now playing in the e-factory. The broadest sense of the term technology is used herein. Technology is defined to mean the application of machinery or scientific principles to performing work. In much of the mass media today, technology is used as a synonym for computer hardware and software. However, in the e-factory, technology must include the innovations in mechanical, electrical, fluidic, optical, electromagnetic, and electronic devices.

Several technology trends have been converging over the last five years to provide the opportunity for quantum leaps in manufacturing productivity. One can speculate that much of the consistent high growth in the economy with the absence of inflation has come about because of the high level of productivity improvement that has occurred on an annual basis for the last seven or eight years. With the advent of the acceptance of the Internet as a means of communicating and conducting business, the potential for continued productivity improvement has been expanded. However, innovations in conducting business because of the Internet will not by themselves create productivity improvements. There must be innovations in the internal operations of manufacturing and service businesses as well.

Overview of the Impact of Technology

The enterprise and supply chain models described in Chapter 1 have been created for the purpose of illustrating how the flow of work, material, and information all contribute to the creation of satisfied customers. Innovation in technology has had beneficial impact on the key business process parameters. In the following, how these have been beneficial and their connection to the e-factory concept are described.

The key technology areas discussed in the following include sensing, actuation, connection, integration, and control. Innovation in these technology areas has made it possible for the business processes that operate within a manufacturing company and within a supply chain of companies to be more flexible, to depend on less inventory, and to have a high rate of throughput.

Technology innovation began centuries ago with simple mechanical systems that began to animate man-made objects. The introduction of electricity accelerated this animation and facilitated the development of electrical sensors, signal conditioning, logic, and actuators. During the last 20 years, microelectronics have steadily grown less expensive, smaller, and more power-efficient. The replacement of hard-wired designs with embedded processors enormously increased their input-output capabilities, allowing more sophisticated user interaction. This new type of two-dimensional user interaction brought with it the attendant problems of the many buttons and displays that demand attention in the digital world.

This trend will explode in the very near future as technology crosses the boundary where it becomes feasible to imbue most of the common (and even disposable) objects in our environment with some degree of sensing, communication, actuation, and processing. Present-day user interface logjams should evaporate as distributed sensors and processors capture our gestures and collaborate to evaluate our intent, while appropriately embedded actuators and displays give us feedback about the real and virtual worlds.

Sensors

Over the last 30 years, there have major breakthroughs in sensor technology due to research in the defense industry as well as in academia and industrial technology centers sponsored by government and business. Sensor technology allows the physical world to be measured electronically. Electronic measurements can then be converted to digital information that then can be used by microprocessors for feedback control, report generation, and environmental monitoring. In the e-factory, in order to achieve speed and flexibility, the ability to rapidly change the state of production on every machine and in every task in the plant is essential. State change control requires that the production state is electronically measurable, which explains the need for intelligent sensors to capture this information. Exhibit 3.1 depicts some of the relationships between sensors and operations

Sensor technology today is intelligent and based on a variety of physical phenomena that includes electromagnetic, infrared, optical, mechanical, and chemical effects. Sensors are designed to sense absolute or relative position, fluidic flow, physical motion, speed and acceleration, temperature, electromagnetic and optical emission or reflection, smoothness, thickness, noise, and

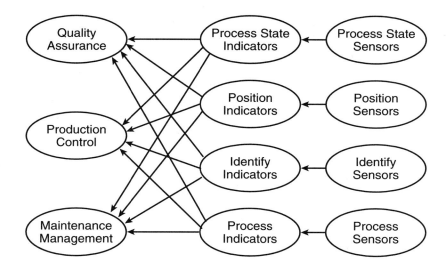

Exhibit 3.1   Relationship of Sensing to Operations

chemical composition. Sensors can be separate components and devices or they can be integrated into solid-state electronics or mechanical structures. Sensors are typically used today to capture data about position, identity, physical state, and environment.

Position sensing is important for automation systems such as pick-and-place robots, welding robots, and material handling systems because there is a need to have physical confirmation that an object that is being handled and is supposed to be placed in a certain location is actually at that location. Automation systems have been applied for many years to only a few types of work. The limit to their application has always been the sensor technology for providing feedback of physical positioning.

Identity sensing is important because it provides absolute confirmation about the location and status of the object being tracked. The identity sensing technology that is the best known to the general public is the laser scanner at the checkout counter at the grocery store. This technology allows a bar code to be read from almost any orientation, thus simplifying the process of identifying the labeled object for the automated invoicing system. Identity sensing includes the sensing of the identity of individual items such as at the grocery store checkout counter, batch sensing for large lot size manufacturing processes, and process sensing for continuous flow production processes.

Physical state sensing is important because when a process is to be speeded up, slowed down, or changed over, the physical state of the machinery and the material involved in the process must be identified so that the proper actions can be taken and the proper controls can be executed.

Physical state sensing is also important for managing the maintenance of production equipment. There is a growing body of knowledge being developed by manufacturing equipment suppliers as well as their users about what are the key indicators of equipment wear and performance tolerances. This information can then be used in maintenance management and process control software to indicate when equipment or processes are about to go out of the control limits required for quality and when they should be repaired or adjusted.

A growing area of application is in material state sensing. This is sensing about not only the location or identity of material, but also what processes have already been performed on the material and what may be next. The objective of this type of sensing is to simplify the cell control or production line control system by allocating some of the responsibility for material, process tracking, and schedule data to the material itself. This concept has evolved out of the same trend in electronics miniaturization that is benefiting personal communications and computing. The idea is that if finished integrated circuitry can be built into a "smart" credit card, this same technology can be used to build a smart identifier label that can be attached to the carrier that is moving material through a production line or attached to the material itself. This smart label would have onboard the necessary communication circuitry, a micro-processor, and memory for storing the product schedule for the material and the production history for the material. This relieves the cell or production line control from having to keep track of every action and measurement for every piece of material moving through the plant while maintaining the capture of detailed real-time information needed for control and quality monitoring. Real-time communication between the machine and the material also reduces the chances of errors in processing material.

Environmental sensing is also a key area of improvement in the application of sensor technology. By being able to sense more reliable and consistently the key parameters used to quantify environmental safety and comfort, the chances of accidents and costly mistakes can be reduced. Environmental sensing includes not only sensing hazardous materials, but also parameters such as temperature, humidity, pressure, particulate count, and fluid flow.

The net effect of all these sensor developments and applications is that computer-based intelligence can be applied at every level of the factory all the way down to the material itself. This allocation of intelligence to everything physical provides the underlying capability necessary for the e-factory.

Connection

With sensors becoming intelligent and because manufacturing equipment is already intelligent, the opportunity to have a real-time controllable e-factory is limited only by the ability to electrically connect all the intelligent devices. And with the recent explosion of cheap personal communication technologies, the factory can become immersed in an electronic ether where all the intelligent dots are connected by electronic communication devices. Communication can be based on wires, fiber, or wireless technology. There are personal computers and microprocessors everywhere within the factory and everyone in the supply chain wants to communicate via the Internet outside the factory. The technologies being developed to support the Internet communications concepts — that is, anyone can communicate with anyone at any level at any time — are now applicable to the factory environment.

Many of the key issues facing decision-makers about connection choices have to do with media, bandwidth, cost, reliability, flexibility, and usefulness. Media issues include choices between copper wires and optical fiber, or between wires and wireless, or some combination of them all. Bandwidth issues have to do with how much and how fast data needs to flow between the intelligent devices. Cost is an essential ingredient because all of the alternatives for media and bandwidth have major cost implications. Reliability is an additional factor because some of the newer media technologies do have reliability issues that have long been solved by the older media technologies.

Flexibility is a new requirement spawned by the needs of the e-factory. The definition of flexibility is a bit fluid itself. However, flexibility usually means being able to communicate from a variety of places, being able to reconfigure the network rapidly, and being able to enter the network from anywhere. The more flexibility required, the more the media balance tips to wireless technologies. The final question is how useful these connection options are and that depends on how well the connection technology fits the requirements of the e-factory. Giving everyone in the factory a cell phone or a palm computer just because it can be done does not mean that benefits immediately follow. Some of the most flexible and most productive factories today operate with handwritten signs and messages on erasable boards in full view of continuous flow manufacturing cell teams with computer keyboards in only a few locations.

The variety of communication media available for application in the e-factory has been growing rapidly over the last ten years. A summary of these media types and their comparative advantages is given in Exhibit 3.2.

Integrating the connection technologies within the control hierarchy and within the business processes is the usually the most important decisions. The classic example is what happens when a company moved from voice mail to e-mail for their primary internal communications. First, everyone had to have a personal computer bought for them. Then there had to be a corporatewide computer network installed. Then there had to be training on how to use the e-mail software. Then the people who were mobile much of the time, such as salespeople and field service staff, had to have portable computers. Then when people got use to using the e-mail system, they found that they could copy an infinite number of people with their e-mail messages. This led to the phenomenon of people receiving hundreds of e-mails a day from people they did not really know, rather than receiving ten voice mails a day from people they did know. The point of this example is that as a business process moves from dependency on one type of connection technology to another, the total cost and impact on the process needs to be understood and planned for in advance.

Connection integration issues in the e-factory include not only selecting the connection media but also designing the network software architecture.

Exhibit 3.2   Connection Media Alternatives

Media Type	Advantages	Disadvantages	e-Factory Requirements
Wire: twisted pair	Inexpensive; easy to install; easy to secure	Low data rate; no mobility; susceptible to radio frequency interference	Wiring; number and location of connections; volume capacity needed; number of digital and analog circuits; network architecture; switch technology and location
Wire: coax	Moderate expense; easy to install; easy to secure; high data rate; not susceptible to radio frequency interference	No mobility	Wiring; number and location of connections and head-ends amplifiers; capacity needed; network architecture; switch technology and location
Wire: optical fiber	Moderate expense; easy to secure; highest data rate; not susceptible to radio frequency interference	No mobility; significant installation care required	Wiring; number and location of connections; capacity needed; network architecture; switch technology and location
Wireless: infrared	Provides mobility, especially in electrically harsh environments often found in industrial settings	Expensive to install; difficult to secure; low data rate; requires line-of-sight; susceptible to signal blockage; many connectable units required	Number of connectable units; types of connectable units; interfaces to connectable units; data capacity needed; network architecture and location of transmitters
Wireless: radio frequency	Provides mobility with a local range that can extend for thousands of feet; inexpensive to install	Difficult to secure; moderate data rate; many connectable units required	Number of connectable units; types of connectable units; interfaces to connectable units; data capacity needed; network architecture and location of transmitters

Exhibit 3.2 Connection Media Alternatives (continued)

Media Type	Advantages	Disadvantages	e-Factory Requirements
Wireless: cellular	Provides mobility with extensive range; inexpensive to install; broad range of commercially available services; can include messages, voice, data, and images	Moderate cost to operate; moderate data rate; many connectable units required	Number of connectable units; types of connectable units; interfaces to connectable units; data capacity needed; network architecture and location of transmitters

The first decision is whether the communication technology will be digital or a combination of analog and digital. Digital is clearly winning out worldwide for most applications, although there may be some legacy equipment or systems issues that require a hybrid. A second decision is what type of communication switching and network protocols to use. This will depend on the communications vision for the e-factory.

Actuation

Most factories generally consist of machines and people who use a variety of electrical, mechanical, and fluidic devices that convert purchased material into final goods. In the factories that are falling under the influence of electronic commerce, there is a rapid evolution from a paradigm of people operating and controlling individual machines or production lines to one of people analyzing data from computer-controlled machines and production systems. As the cost of computer control technology continues to decrease while its capability increases, the search for new ways to improve the performance and cost effectiveness of all the technology in the plant also continues.

One of the areas where there is a major revolution just starting is in actuation technology. Actuation is the conversion of a signal to move into a mechanical movement. The evolution of actuation technology begins with the water power systems of the 18th and 19th centuries where the kinetic energy of running water was converted to mechanical power and delivered by belts to individual machines in a factory. The advent of electric power in the 20th century brought the convenience and flexibility of being able to transport electric energy anywhere and convert it to mechanical power with pneumatics, hydraulics, and electric motors. Fluidic systems, whether they be pneumatic or hydraulic, actuate mechanical motion by converting pressure on a fluid acting against a piston. By changing the pressure, the piston moves and mechanical motion is achieved. Electrical actuation has traditionally been based on converting electrical energy into rotating or linear mechanical motion.

Hydraulic systems have been popular for many decades for applications that include lifts and brakes because they offer big leverage for power transmission and wide ranges of acceleration and deceleration of motion. They have a long history of successful operation and application. There are, however, several factors that are causing the future of hydraulic systems to be challenged by electrically actuated systems.

One issue is environmental. Hydraulic systems are fluidic and require high-quality plumbing for them to be completely free of leakage or spillage of hydraulic fluid. Unfortunately, cleaning up of hydraulic fluid is one of the maintenance functions that goes hand-in-hand with hydraulic systems.

A second issue is that of cost. Hydraulic systems have been cost-effective for many years because of the low cost of the petrochemicals used as hydraulic fluid and because of the low cost of electrical energy. A hydraulic system operates through the transfer of hydraulic pressure into mechanical motion. However, hydraulic pressure is maintained by an electric pump in most systems. The electric pump has to operate almost continuously to maintain the necessary pressure in the hydraulic system, even when there is no motion being created.

Thus, the obvious question is how much efficiency is lost and cost increased between the electric motor keeping pressure on the hydraulic pump and the mechanical motion. There is a trend now to replace electrically assisted hydraulic systems with electric motor actuation. Using this type of system, electrical energy is consumed only when there is a need for linear mechanical motion. In addition, all the environmental issues are eliminated.

In addition to electric motor rotary to mechanical linear actuation, there is a new area of actuation beginning to appear using linear motor technology. The design of a linear motor is equivalent to taking the cylindrical arrangement of magnets found in the classical electric motor structure and laying out the magnets on a flat surface. The rotor of the motor then reacts to the magnets that are laid out in a straight line and the motion that results is "linear." This direct conversion of electromagnetic energy to linear motion is becoming more attractive for many applications that previously would have used electrical to mechanical linear motion.

The reason that the technologies associated with actuation are being reviewed here is that electrical actuation provides an improved level of performance and flexibility that is amenable to computer control. The concept of control by wire that evolved first in the aerospace industry is now being applied to the automotive industry and in manufacturing automation systems in general. In fact, a robot is really nothing more than a system of control by wire actuators, sensors, and computers. Material handling automation consists of the same technology. All the key performance factors and criteria for the e-factory require the flexibility, power, and speed of electrically actuated systems. As a result, they are becoming a key part of the e-factory.

Control

Just as life seems more complex today, the systems and business processes in the e-factory are more complex than those of the past. They are complex because they involve many subsystems and components, many points for decision-making, many variables to control, and they are under greater stress to be fast, accurate, and flexible. Over the last few years, there has been much thought put into the concept of complexity and how to control it or survive it. There is, in fact, a body of work called Complexity Theory that is the successor to Chaos Theory. Chaos Theory was developed by psychologists to explain the observation that certain (and maybe all) large natural systems could appear to be operating in a stable mode and then some small event would propel the entire system into chaos. These models were used to describe the behavior of lemming herds when they would suddenly propel themselves as a group off a cliff for no discernable reason. However, Chaos Theory tended to be too depressing (any larger system was deemed susceptible to it) and of little value to those who want to solve problems. It did lead to the development of Complexity Theory.

Complexity Theory says that complex systems can be controlled and operate successfully without falling victim to the lemming behavior. The answer from Complexity Theory is that complex systems should be organized into simpler "affinity" groups that operate with simple rules. The rules for each group are based on the goal of optimizing its operation in its local environment. The answer to managing a complex system is to break down the complex system into simpler subsystems that can be more easily controlled individually. The total system has the best chance of being stable or under control.

Process and control simplification has been the basic concept behind much of the work in factory floor process improvement methodologies and in control systems, and has been contrary to the work behind integrated enterprise systems. The concept and benefit of cellular manufacturing can also be explained by Complexity Theory. Thus, much of what must be determined when building the roadmap for the e-factory is how to decompose the complex system into controllable simple cells and subsystems.

The control technology for the e-factory includes electronic hardware, microprocessors, software, sensors, and actuators. This control technology must be implemented in a control hierarchy that is consistent with the capabilities of the control technology and is designed to implement a control strategy for the entire plant. The control hierarchy used herein begins at the enterprise level and is broken into four levels down to the individual machine and material levels. A generic example of this control hierarchy is illustrated in Exhibit 3.3, which is identical to Exhibit 1.5 in Chapter 1 but repeated here for convenience.

There has been much innovation in factory control systems, primarily as a result of the innovations in computer technology. The modern factory floor and production line control systems are designed to sense and respond to reality while executing a control strategy. The key to this is the use of feedback

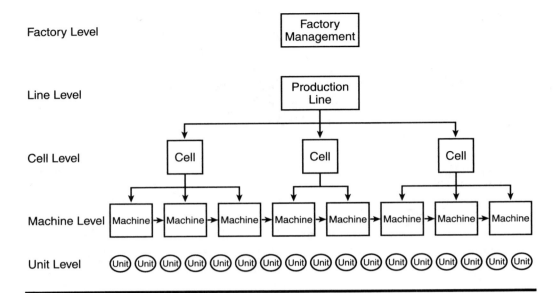

Exhibit 3.3   e-Factory Control Hierarchy

control concepts. Feedback control concepts were developed early in the last century as a way to assist in the operation of steel mills, petrochemical plants, high-performance aircraft, missile systems, and other precision-guided weapon systems. The theory proved to be so powerful that there have been attempts to apply it to modeling and controlling economic systems as well as biological systems. However, the application to factory systems has been a bit long in coming, primarily because of the cost of and the lack of effective control software and hardware technology. There were a few attempts at fully automating a manufacturing environment, but the number of "lights-out" factories that operate on a regular basis is very small.

The concept of feedback control is simple. If a self-propelled vehicle were designed to follow a painted strip on the floor with an optical sensor, the control software for the vehicle would be monitoring the distance between the painted strip and the center of the vehicle. This distance represents the position error for the vehicle. When the position error exceeds a predetermined limit, say ten centimeters, the control software sends a signal to the steering system to turn the wheels in the opposite direction from which the error is increasing. Thus, if the vehicle is beginning to drift to the left of the strip, the steering signal is to turn the vehicle to the right until the measured position error falls back within the limits. The "feedback" of the position error from the optical sensor to the control computer allows the control computer to calculate a correction signal that brings the vehicle back to the strip. This type of feedback control is essential for high-volume production processes and all chemical manufacturing processes. It is now becoming available for virtually all types of manufacturing and is being designed into most manufacturing equipment systems. A depiction of feedback control in the e-factory control hierarchy is given in Exhibit 3.4.

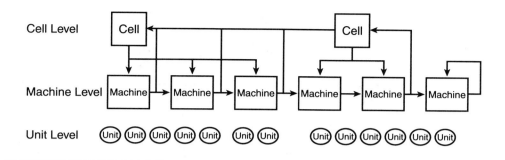

Exhibit 3.4   Feedback Control Concepts

Most plants in the past have tended to operate in an open-loop fashion for long periods of time. A production schedule would be produced by a manufacturing resource planning (MRP) system for a period of one week for example (or in some cases for one month) by estimating what product needed to be produced based on sales forecast or trends. At the end of that one-week production schedule, a new MRP compilation is performed that produces a new production schedule for the next production period. The new production is based on a new forecast less what may have actually been sold or shipped during the last planning period. Thus, during the production schedule period of one week, the plant is running open-loop. There is no feedback on what customers actually want or what is actually being shipped. This is compounded when the total lead-time from supplier to finished product for products is measured in weeks or months. An MRP system has to use forecast horizons of months to accommodate the lead-times of suppliers. This tends to create inventory levels that get out of control because there could be several months of production of a certain product in the pipeline without any feedback on what the market really demands.

The e-factory environment requires short cycle times for producing products and real-time feedback on how well demand is being satisfied by production. The Dell Corporation revolutionized the personal computer industry by being able to take an order over the phone or the Internet for a computer that was configured by the customer while on-line and then have the product shipped to the customer within two days. The production process was designed to be able to build to order a computer within a cycle time of a few minutes and to have the order shipped within 24 hours. This type of "real-time" control requires that the production process be designed to be this flexible, that the materials management system be designed to respond this fast to demand, and that suppliers be able to respond this fast to the production process.

From control theory, it is well-established that for a process to be controllable, certain key parameters need to be "observable." Observability in simple terms means that if the data that describes the temperature of a critical process cannot be measured by a sensor or instrument, then there can be no control algorithm developed to change the process to keep the temperature within the acceptable range. Sensors are critically important to the future of the e-factory because they provide the means of observability from the physical world to the digital world.

If all the key parameters are observable, then it is necessary to ensure that the process is controllable. Controllability means that the process is designed so that rapid adjustments can be made to bring a key parameter back into acceptable limits once it has been "observed" that the key parameter is approaching or has already exceeded those limits. If a gas-heated furnace is supposed to operate at 500°F within a range of ±2°F and a temperature sensor observes that the temperature has reached 503°F, then control software should reduce the flow of gas or oxygen to the furnace burners until the temperature reading falls back into the range of 498 to 502°F.

In general, the capability, availability, and importance of manufacturing control software has increased over the last ten years. During the 1980s, most companies installed manufacturing resource planning software or cost accounting software and let the people running the production shop floor and manufacturing scheduling look after their own problems. With the need for faster information flow, tighter control, and the need to feed other manufacturing plants based on JIT systems, the need for software that can provide real-time control continues to increase. Most enterprise resource planning (ERP) systems and manufacturing resource planning (MRP) systems are designed to provide planning tools and activity tracking, but have not addressed the real-time control needs of the factory floor. Organizations such as IMS (Intelligent Manufacturing Systems Consortium) and MESA International (Manufacturing Execution Systems Association) have grown out of the needs of manufacturers and computer hardware and computer software suppliers to set standards for the next generation of manufacturing execution and control systems.

The challenge in advancing e-factory automation and control lies primarily in overcoming the difficulty in integrating the wide variety of intelligent machines and people operating in the factory. There have been a few standards developed over the years (e.g., MAP, GEM/SECS, etc.) for integrating equipment into an e-factory network. However, more revolutionary approaches are being taken to address the control and integration issues.

Systems Integration

Integration in the factory environment means connecting all the intelligent agents (human and machine) in the environment and providing them with the necessary sensor technology and control software so that the factory can operate efficiently, with speed, and with flexibility. In many ways, the control software cannot operate successfully in the e-factory unless it is also acting as the integration medium. Total electronic integration requires computer control. The maximum benefit of computer control will not be achieved unless there is total electronic integration.

In reality, the control software developed over the past ten years has more than kept up with the integrating technology. However, technology is at a point where the control software and the concepts it is built on are now a limiting factor to the degree of integration that can be achieved in a factory environment. As a result, there are several global development efforts underway to address these issues. These global development efforts will likely begin to show up in commercially available software and systems in the near future.

For example, the IMS Consortium has been engaged in factory floor real-time control research over the last five years. Membership of the IMS includes the United States National Institute of Standards and Technology (NIST), the National Center for Manufacturing Science, Rockwell International, United Technologies, Fujitsu Fanuc, Hitachi, Toshiba, and Yaskawa Electric, plus many other European academic and industrial organizations. One of the fruits of their research has been in an area called "holonics." The term "holonics" is about the design of "holons," which is derived from an ancient Greek word that means a self-contained object. The term was coined in Europe about 30 years ago for social systems, and more recently has been applied to manufacturing systems.

Over ten years ago, a similar project was initiated by SEMATECH, an U.S.-based semiconductor industry consortium composed of semiconductor manufacturing equipment makers. This consortium developed a computer integrated manufacturing (CIM) framework that was based on object modeling for semiconductor manufacturing. Included in the CIM framework is an object model for machines. There is a sister organization called SEMI composed of semiconductor equipment manufacturers, which has also developed an object model for machine interfaces called the Object Based Equipment Model (OBEM). These models are ambitious attempts to standardize the control components within a standard model for a factory in general and equipment in particular. The CIM framework for equipment actually attempts to identify subcomponents within a machine that are specialized for processing, measurement, and material handling. It also has the concept of ports where material enters and exits the machine.

A key element of these developments is object-oriented technology (OOT). OOT has been touted for many years as a good way to modularize software components that can then be reused efficiently. It has taken hold in programming languages (C++, Java) and operating systems (Microsoft Windows), but it is only recently that it is beginning to have any influence on business process and manufacturing models.

An important objective of e-factory control and integration software should be to reduce the cost and time of configuration and integration efforts. Using object-oriented concepts in e-factory software can provide "plug-and-play" capabilities for all of the e-factory components so that they can be plugged into a system and automatically configure themselves to work correctly. This is a very useful objective for the next generation of equipment and business process interfaces. It should be possible to plug a new piece of equipment into a manufacturing system and have it identify itself automatically to the system and become productive immediately.

There is more to running an e-factory than an integrated factory floor control system. Much more important is higher level integration between other e-factory systems and business processes such as product realization, enterprise resource planning, and production scheduling. The same plug-and-play concept should apply to these functions. An equipment object should have all the attributes and methods necessary to support these functions. This could mean, for example, that equipment should be able to describe its processing capabilities and even simulate them for planning.

One of the areas where research is advancing quickly is in object-oriented interfaces. These types of interfaces are based on object models and can be implemented using Object Request Brokers (ORB) rather than using message systems which the first MES packages used. There is a bit of competition between international standards organizations and certain corporations about how this technology will be defined and controlled.

A key problem in defining e-factory machine interfaces is the distinction between simple and compound machines. The concept of a simple machine is a device that can perform one processing operation on one group of material at a time. Examples include a machining center, an ion implanter, or a wafer lapping machine. These machines have a well-specified work cycle that works on one part at a time, or a batch of parts serially, or a batch of parts together. It has a well-defined cycle time, capacity, and state (up or down).

However, it is normal to combine several simple machines into a machining system or cell, such as a cluster tool or a flexible manufacturing system. It is more difficult to define the capacity or state of a machining system or cell. To have a flexible plug-and-play environment, the e-factory must have a configurable environment for simple and compound modules or cells. The e-factory environment would be further simplified if it were made up of recursive machine objects.

Recursion of machine objects means that a group of machine objects can be combined into a cell that is also a machine object. This process can be repeated to combine cell machine objects into an e-factory machine object. Similarly, e-factory machine objects could be combined into an enterprise or a supply chain machine object. The key is to generalize some of the characteristics of a machine object so that they can apply to a group of machine objects.

Industry Integration

In parallel with systems integration developments in the e-factory over the last few years, there has been ongoing integration and transformation of the industries that provide the manufacturing equipment, control systems, and application software to the e-factory environment. This transformation is being driven by the need to improve the effectiveness and efficiency of the factory automation being introduced to the e-factory. In many ways, this is a result of the ongoing trend among manufacturing companies to outsource as much of the design and operation of a manufacturing process as possible and minimize the overhead burden of maintaining an in-house manufacturing process design team. This transformation has resulted in a stream of mergers and acquisitions of hardware and software companies that evolved to produce specific solutions or capabilities. The ultimate strategy of the companies spearheading these mergers is to be able to provide an e-factory "solution" rather than just a collection of components that need to be assembled into a solution by the customer. This trend of integrating point solutions into process line or cellular solutions is emerging from two different directions, as illustrated in Exhibit 3.5.

One strategic direction of industry integration is occurring from the top of the control architecture down through the lower levels; it is based on offering solution sets of high-level control software and computer hardware and extends down to the lower control levels of device control computers and sensors. Companies that are pursuing this strategic direction include Rockwell Automation, Invensys, Hitachi, and Brooks Automation.

The other strategic direction of integration is from the bottom of the control architecture up. The bottom of the control architecture begins with the machinery and equipment that handles and converts material as it becomes a finished product. This includes linear motion systems, control by wire electric actuators, fluidic actuators, process-specific production equipment, and material handling equipment. The trend in intelligent machines and machine objects described above is stimulating this integration trend. The benefit of this type of integration is that real-time control can be implemented throughout the entire production process without being dependent on stand-alone generic computers. Companies that are pursuing this integration include Danaher, Parker Hannifin, Mannesman Rexroth, Dover, and Applied Materials.

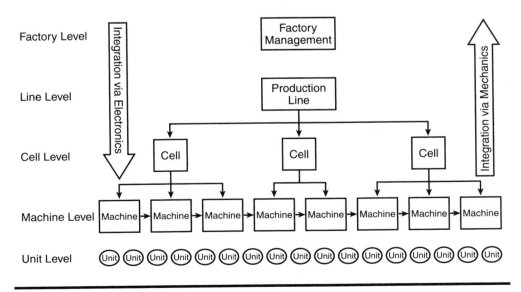

Exhibit 3.5   Integration Directions

Performance Measurement

Real-time control of the e-factory requires not only the right technology but also the right performance measurement strategy. Performance measurement is necessary for the production line, plant, and enterprise level of controls.

The best performance measurement systems use activity-based performance indicators. An activity-based architecture supports process management rather than departmental management, and provides for decisions based on time and actions rather than on budgets and forecasts. Using activity-based cost measures makes it easier to understand the impacts that certain decisions make on profitability as well as customer satisfaction. Using activity-based measures also generates information that can be used for feedback control. Activity-based cost modeling performance measures of quality, reliability, and productivity can be converted into monetary units. Exhibit 3.6 illustrates the differences between activity-based costing and departmental costing.

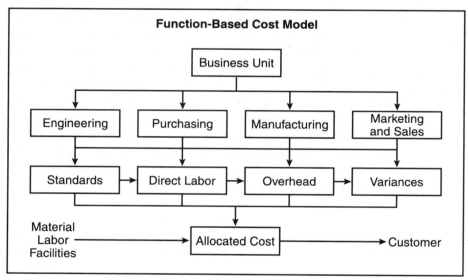

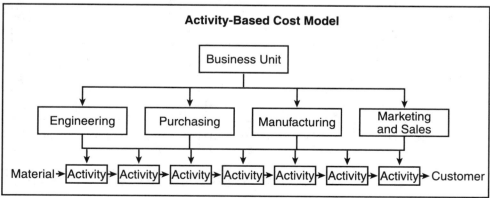

Exhibit 3.6   Function-Based versus Activity-Based Cost Modeling

There are typically four levels of activity-based costing. The first level is the unit level. Unit level costs are associated with the actual creation of a unit of product and include the direct and indirect activities that contribute to the creation of the unit. All resources (human, machine, utilities, materials) are included in the unit level. The second level is the batch level. These are the costs associated with the setup and teardown of the business process for each batch or run. Included in batch level are the direct and indirect activities that contribute to the setup of the business process. The third level is the product sustaining level. Product sustaining costs are associated with the development of the product and the sustaining of product design or manufacturing process. This level includes pre-production and post-production activities that contribute to the design and creation of the unit. The fourth level is the facilities level. The facilities level of costs includes costs associated with the facility and other very general and very indirect activities. It includes overhead activities not associated with the production of the unit as well as overhead allocation.

The key performance measures should include throughput, quality, cost, productivity, capacity, and availability. Throughput measures include output per unit time of material through the process, the cycle time of the process, variety of throughput, and throughput lot size. Quality measures of performance include first-pass yield, rework levels, and process control parameters. Quality measures should be captured at all levels of the control hierarchy, at each step in the process, and at each node in the supply chain. Productivity measures of performance include the productivity of capital equipment resources, human resources, and facilities. Availability measures of performance include identifying the availability of capital assets, human resources, current production capacity, future production capacity, and inventory. The usefulness of availability measures is in determining how much scaling up or down can be done, whether additional resources will be needed, and whether existing resources can be redeployed to new applications.

Summary

Technology, in its broadest definition, is acting as a facilitating agent and catalyst for the e-factory. The technologies that offer the greatest promise through innovation include sensing, actuation, connection, integration, and control. They are making the e-factory a reality by making it possible to have short cycle times, flexibility in process changes, rapid response to changes in demand, and minimal inventories. Identifying how to best use technology in such a way as to make the new business processes operate successfully is a key part of developing the e-factory roadmap.

Chapter 4

The Role of Process in the e-Factory

Purpose of This Chapter

Every business enterprise is a combination of activities that results in the creation of value in the form of a finished good or provided service. All the activities can be grouped into one or more major business processes. Each business process operates as an integrated system of people, capital, and technology. It is the structure of the business process and how people, capital, and technology are organized that makes the difference in the performance of the enterprise. Technology depends on process and amplifies process. The purpose of this chapter is to describe the role of the business processes that are critical to the operation and evolution of the e-factory.

Why Process Is Important

Process is important in the e-factory. In fact, the e-factory is all about process and how technology is changing it, supporting it, and accelerating it. All of the big changes that are taking place in the manufacturing environment are happening to processes. The good news is that most of what has been learned about best process practices and applied over the last decade continues to be useful. In fact, much of the technology that is now available for the e-factory would be wasted without applying the process design concepts and best practices that have been developed by a variety of companies and institutions.

The long evolution of best practices in business process design and operation is converging on a set of principles that is creating a consensus approach to business process configuration. This common set of principles represents a merger of the thinking and practice that has evolved from under the banner of several names that include Total Quality Management, Concurrent Engineering, Continuous Flow Manufacturing, Business Process Reengineering, and Lean Manufacturing. The common set of process principles include the following:

- A business enterprise should be viewed as a set of critical business processes.
- A business process is a repetitive flow of activities that results in a satisfied customer (through such events as developing and introducing a new product, or delivering finished goods in response to purchase orders).
- Each activity consumes resources.
- Each resource has a cost and key cost driving parameter.
- Business processes should be managed by process teams.
- Process teams should be cross-functional and consist of people from appropriate functions.
- There are skill sets or activities that are core competencies for the business enterprise that must be maintained and expanded.
- There are skill sets or activities that are not core and therefore may be outsourced or purchased from other companies.

If the e-factory for which a roadmap is being developed is a green field facility (i.e., a new or renovated building), then all new processes will have to be developed. If the e-factory roadmap is for an existing plant that is being transformed, then the existing processes will have to be reengineered and transformed as well. The approaches to both new process design and transformation of an existing process are similar in most respects.

It is highly probable that the definition of a business process will cross over the boundaries of several functional organizations within the enterprise. The approach for designing and implementing improvements must be focused on the process rather than the functional organizational structure. Process improvement requires consideration of all types of resources, including human, capital equipment, information, and consumed materials. Most methodologies for process design and improvement fall into two categories: continuous and discontinuous.

The process design concept of continuous improvement is best applied to existing processes because it encourages a mentality and work ethic within the process team that causes any opportunities for improvement in specific activities to be acted upon as quickly as they are identified. Continuous improvement as a philosophy has been an important part of virtually every form of quality training ever offered.

However, incremental improvements can eventually reach a point of diminishing returns and usually are not appropriate for new process designs. One way to break through the barrier of diminishing incrementalism is to do something bold or dramatic. Changing the entire business process rather than just changing specific activities is just such a bold or dramatic approach. It is this "start with a clean sheet of paper and challenge everything you ever believed" paradigm shifting approach that is appropriate for truly transformational process situations.

Ultimately, it is the process that produces the results at the internal business process level, at the external business process level, and at the supply chain level. Good technology will amplify good processes and bad processes. If the process is not efficient, technology will make matters worse. If the process is efficient, the level of technology needed may be less than expected.

The cost of technology is determined by process capability. Processes designed for real-time feedback control require extension technology. Extremely simple processes require less costly technology. Processes that have not been designed, but which have evolved, are inherently complicated and usually operating on the edge of chaos.

Exhibit 4.1 illustrates the concept of internal and external business processes and how they relate to the key business processes in the supply chain. Understanding the difference is important because the two types of processes are owned by different entities. Internal processes are owned by the enterprise. External processes are jointly owned by customers and suppliers.

The processes that are internal to an enterprise include product realization, order fulfillment, order capture or customer relationship management, and administrative support processes. These processes represent the heart of enterprise operation and the bulk of where value is added in the enterprise.

The business processes that operate externally to an enterprise are those that are owned by an entity external to the enterprise but which have significant interaction with the internal business processes. External processes often include purchasing, supply chain planning, logistics, and transaction settlement. However, if a company has undergone supply chain fission, almost any internal process could become an external process. In the following, both internal and external business processes are discussed.

Internal Business Processes

The internal business processes are designed, owned, and operated by the e-factory enterprise. Typically, the internal business processes include some or all of the following:

1. order fulfillment process
2. product realization process
3. customer relationship management process
4. support processes

In the following, each of these internal business processes is examined and their impact on and importance in the e-factory discussed.

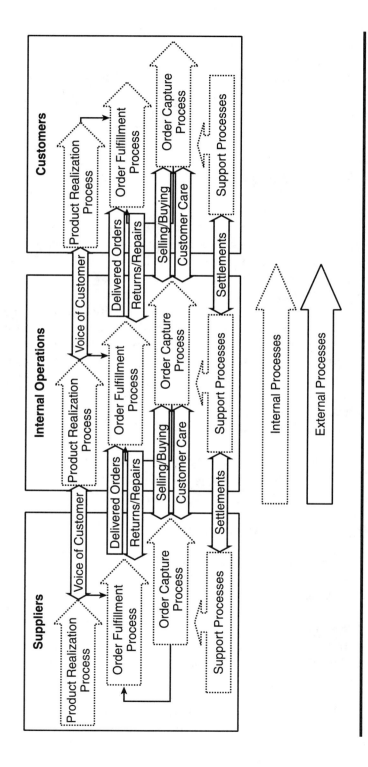

Exhibit 4.1 Internal and External Processes

The Order Fulfillment Process

The order fulfillment process converts orders into delivered goods and services. Order fulfillment begins with the receipt of an order and ends with the delivery of the ordered products to the customer. The order fulfillment process can vary widely from company to company depending on how the company adds value to a supply chain. The activities involved in order fulfillment include:

1. production planning
2. materials planning
3. purchasing
4. inventory management
5. warehouse management
6. production control

Order fulfillment includes the flow of work, material, and information. The flow of information begins with the creation and acceptance of an order. The information that must be managed includes order status, production status, process planning, maintenance information, and schedules. The order fulfillment process interacts with other internal enterprise processes such as product realization, customer relationship management, and the administrative support processes.

Production Planning

Production planning involves the analysis of production capacity and the demand for products to determine how to allocate production capacity to the production of forecasted or demanded products. The complexity of the production planning activity depends on the number and diversity of products that can be produced within the facility, the frequency with which the production plan needs to be changed, the capability of the production facility, and the quality of the feedback on what the facility can execute.

Traditional planning horizons vary by industry and by manufacturing strategy. Horizons in the automobile industry range from six months to seven days. In some electronics industries, planning horizons range from seven days to one day. The advent of e-business is causing the e-factory to have a planning horizon so short — on the order of a few hours — that production planning as it has been traditionally known is being abolished. The new e-factory requirement is that a production process should be able to produce product on demand, that is, build-to-order. The production planning challenge then becomes one of making sure that the production operation has the capability to build-to-order.

Thus, the production lines have to be designed to be able to build a wide variety of products or product configurations with minimal or no setup time. This usually also requires that the products be designed so that they can all be built on the same production line with minimal setup. The third key element of this is that the factory floor control system must be designed so that it can carry, store, and process the necessary information and instructions. Thus, the design of the product, the production line, and the control system must be created at the same time and with one set of objectives in mind.

Materials Planning

Materials planning is the activity of defining the type, quantity, and delivery schedule of the material needed for production during a given time horizon. The key drivers in determining the nature of the materials planning activity are the process cycle times for production and suppliers. If the lead-time or cycle time for suppliers is larger than the production cycle time for the order fulfillment process, the typical approach to materials planning is ask for a forecast for product sales that looks farther into the future than the supplier's and internal process cycle times. Using these forecasts, materials planners or the material resource planning (MRP) software generate a list of the required materials, compare the list to material in inventory, and then generate a list of the material to be ordered from suppliers.

Materials planning plays a key role in the order fulfillment process because it is this function that determines how much risk is being taken or how much uncertainty exists for the enterprise's investment in inventory. The farther into the future or the longer the planning horizon, the greater the risk that the wrong products will be built, that there will be excess inventory, and, therefore, greater risk of financial loss or excessive cost.

While MRP software has been in existence for over 20 years and has helped in automating the selection and decision-making process about what should be ordered, it is nothing more than a calculating tool. The information trends in the e-factory and in the supply chain are creating situations where information about product sales and material needs is becoming more readily available. The cycle time trends are creating situations where production cycle times are shrinking to minutes and lead-times from suppliers are shrinking to days or hours. Given these phenomena, materials planning becomes less a forecasting and calculating activity and more an information processing and communication activity.

In the e-factory, materials planning changes dramatically and, in reality, is being transformed into a different function: demand communication. The ultimate materials requirements activity is a virtual kanban. When the supplier response cycle time is equal to or less than the virtual kanban size, then when the production process indicates it needs new materials, it pulls it automatically from the supplier via electronic signals.

Purchasing

The purchasing function includes the activities of identifying new suppliers from whom to buy, negotiation of the buying agreements, and the execution of purchases required by manufacturing. In many companies, purchasing is a separate function. However, there is a growing trend for purchasing to be part of the important decision-making teams within the enterprise, such as new product realization and operations management.

The role of purchasing is one of the most critical because it is at this point where all spending is done. For most manufacturing companies, purchased goods and services represent from 50 percent to 90 percent of the cost of goods sold. It is clear that any opportunity to reduce the cost of purchased goods and services, either through price reductions or through more efficient use of goods and services, offers great leverage in terms of profit improvement. Thus, any tool, method, or technology that offers a way to achieve these benefits is worthy of consideration.

One of the most dramatic trends in e-factory technologies over the last two years has been in the surge of interest in software that helps automate the purchasing function and helps facilitate reductions in money spent on purchased goods and services. Software products are now available that offer the capability of putting all purchasing activities on-line, of enforcing negotiated buying agreements on-line, and of distributing the tasks associated with executing purchasing across a broader set of people and thereby eliminating non-value-adding activities.

There are also new on-line services becoming available that support auctions and negotiations that are often part of the purchasing process. A further trend is e-markets that are envisioned to replace physical markets and exchanges such as the stock exchange, the commodity exchange, or the used equipment market.

With all these new technology and on-line service developments, one thing is very clear. The process of purchasing must be changed radically for these new capabilities to be effectively used and for them not to become a significant risk to the operation of the other internal business processes.

In the future, the role of purchasing will be even more important because of the instant access to new suppler environments, e-markets and exchanges, and buying environments. Because of this new technology and these new tactical and strategic options, the purchasing decision-making process will become more critical and more time sensitive. There will be a need to process and analyze more data quickly and thoroughly.

Inventory Management

Inventory management refers to the control of the location and movement of inventory in the production process. There are generally four areas in the order fulfillment process where inventory is managed: (1) incoming material, (2) work in process, (3) finished goods, and (4) warehouse or distribution center. The management of inventory in a warehouse or distribution center is covered in the next chapter section.

Incoming material should only be stored for as short of a period of time as possible. Some industries that use large quantities of nonperishable raw materials, the lot size or timeliness of incoming material may not be an issue. For those industries that buy finished or semi-finished parts or subassemblies, incoming material storage should only be for the purposes of queuing up for production lines or cells.

Work-in-process inventory represents the material that exists between the initial and final production steps. The size of the work-in-process inventory is directly proportional to the length of the cycle time of the production process and to the lot size.

Final goods inventory is the inventory that exists after the final production steps. Finished goods are usually either shipped directly to customers or stored in a warehouse to wait for an eventual sale.

Best practices in e-factory processes dictate that the need for inventory management in three of the four areas be eliminated. Whatever inventory exists in a production process is only a small temporary queue that is rapidly emptied by shipment directly to customers for orders. The concept of build-to-order implies there is essentially no finished goods inventory. The concept of vendor-managed inventory implies that whatever incoming material is managed by the supplier of that inventory and is not an issue for the owner of the order fulfillment process. The concept of cellular and continuous flow manufacturing implies that there is only one cycle's worth of inventory in work-in-process at any point in time. In the e-factory, inventory management should become an insignificant activity that essentially becomes an element of material handling and production control.

Warehouse Management

Warehouse management is the management of the finished goods inventory that is stored in a warehouse for order fulfillment and final shipment to retail stores or final customers.

If the warehouse is a distribution center, shipments are usually bulk quantities that are sent to retail stores to replenish stocking levels. However, with e-business short-circuiting the retail stores, shipments out of distribution centers are fulfilling single orders of mixes of small quantities of items being sent directly to the end user.

Thus, the distribution center is becoming both a warehouse for bulk shipments to retail stores and an order fulfillment center supporting on-line customers. The result is that management of the inventory in a warehouse takes on the characteristics of managing the stock in a retail store. This is a significant change in the business process of warehouse management, and is causing the software that has traditionally been used to manage the warehouse to become obsolete and forcing changes in the material handling systems and in the layout of the warehouse itself. The warehouse in the era of e-business is, in essence, an e-factory because it must respond to the same speed issues and order fulfillment requirements.

Production Control

Production line control refers to the control of the machines that form the cells of the production line and the flow of material through those machines. There is usually a control hierarchy with several levels of control in a control system. Production controls have several purposes. One purpose is to ensure that the machines are performing the appropriate functions per schedule and that their operating parameters are within acceptable tolerances. A second purpose is to ensure that material is where it is supposed to be and in the state for which it is scheduled. Another purpose is to ensure that the people involved in the process are working effectively and are provided with the information needed for their jobs. Another purpose is to collect data for feedback on errors and on quality problems. The ultimate definition of process quality is to minimize the variation of a process. Production control systems are ultimately designed to serve that purpose.

The evolution of control technology has been driven by advances in microprocessor technology and application software. Current technology has made it possible to put electronic intelligence that can communicate via computer network into virtually every machine and even into material or material carriers. If the technology is used properly, quality can be virtually guaranteed, processes on the factory floor can be reconfigured in real-time, and lot sizes can be adapted instantaneously.

The expectation is that the power and capability of production control systems will continue to improve. The result will be more automation and more computer-controlled systems in a greater number of companies and industries. The need for the flexibility, speed, and quality necessary to create the build-to-order e-factory can only be satisfied by a powerful production control system.

The Product Realization Process

The product realization process (PRP) is the process that converts market needs into manufacturable goods. Product realization begins with the identification of a need in the marketplace that is being unfilled, and completes with the introduction into the order fulfillment process of a manufacturable product. The activities involved in product realization include:

1. product creation
2. product market
3. product timing

The product realization process interacts with other internal enterprise processes such as order fulfillment, customer relationship management, and the administrative support processes.

There is little doubt that time is an important factor for manufacturing companies worldwide. Getting products to market fast is viewed as a key ingredient to commercial success today for many industries. However, speed in product realization is not by itself the single most important factor. If a product gets to market either too late or too early, it can be a failure; but if it is not what the market wanted, then the result is certain to be a failure. For the e-factory, it is important to know what market it is serving and what that market wants, and then to get product to that market fast.

For a manufacturing company, the single most important business process is getting the company's products into the hands of customers before or instead of a competitor's products, and to do it profitably. If this process takes too long, is too costly, results in a poor quality product, does not include the features needed, or does not do all of these things as well as the competitors, then the business enterprise itself is at risk.

A manufacturing company must respond to the needs of the marketplace. Markets are constantly changing because of the changing behavior or desires of the individual buyers in the market or because of the changing array of alternatives from which buyers can choose. The criteria by which customers make decisions are changing and are affected in many ways by the companies that are supplying product to the customers.

A manufacturing company must respond to the activities of the competition. As the markets change, the changes are either being caused by the competition or by other fundamentals in the behavior of the customer base. Watching the competition is another way of finding out if the market is changing and who is changing it.

A manufacturing company must respond to new opportunities in the market. In an economic environment where most markets are now global with much competition, many times the biggest growth opportunities are in market niches that do not yet exist but that can be influenced or stimulated by the companies that could serve that market.

Product Creation

The right product is a product that is successful in the marketplace and that satisfies a market need. It is *not* the product that engineering says is right, or that marketing says is right, or that manufacturing says is right, or that the CEO says is right. Rightness is defined by the customer and the market.

The definition of success for a product or product line can be any one or all of the following:

- The product generated an acceptable operational profit margin (e.g., greater than 35 percent).
- The product captured an acceptable market share (e.g., #1 share, #2 share, greater than 15 percent share, etc.).
- The product generated increased sales (e.g., greater than 20 percent of the total business unit).

- The product performed as specified (e.g., fewer than 3 percent failures in the field).
- The product had an acceptable lifetime (e.g., greater than 3 years).

Once an initial definition and measure of success has been established, the next issue is what factors determine success and this concept of "rightness." There are a variety of factors that can determine that a product is "right." These include:

- *It satisfies the customer's wants or needs.* Customer satisfaction is the ultimate definition of quality. Satisfaction is defined as meeting an expectation or need. The term "want" is used interchangeably with "need" in this case because the difference is based on whether the requirement is self-generated (want) or imposed by the environment (need).
- *It has the right price.* Price is always an issue in one way or another. For products at the low end of the scale or that are viewed as commodities, the right price is the lowest price. For products at the high end, high price is viewed as a differentiating factor. Luxury items are identified as exclusive because they are attainable only by a select few; the barrier to attainability is price. Thus, the perception of an item being a luxury is driven in large part by its price relative to other similar items.
- *It has the right performance.* The term "performance" is often used in product advertising and is generally the term used as the ultimate comparison metric. Does a laundry product make clothes brighter? Does a sports car accelerate faster? Does a stereo produce purer sounds? These are all questions that imply performance as the ultimate measure of success.
- *It appeals to the customer at the right time.* Timing is important because it represents the matching of the customer's awareness of his needs and the ability of a product to meet those needs. There are numerous examples of products that got to market too early or too late. A company cannot afford to be too late to market because the competition will not wait. A company cannot afford to be too early to market because no one will buy if they cannot understand its value.
- *It is better than the competing products.* Ultimately, a customer who does any kind of due diligence will compare competing products on some basis to come to a buying decision. A customer inherently wants to buy the best product, and best implies that the product chosen is better than the rest of the competition.

Thus, one of the key requirements of a successful PRP is that it ensures that there is capture of a clear definition of the customer's expectations, needs, and wants. This is referred to as the "voice of the customer."

Product Market

Understanding the market means defining its members, its boundaries, its driving forces, and the companies competing in the market. Market descriptions include its size, its growth history, and its growth forecasts. The members of a market can be characterized by parameters such as age group, gender, economic levels, geography, religious preferences, technology, government regulation, and job/profession group. These same parameters can define markets boundaries. The driving forces can be price, technology, government regulation, style, demographic changes, and environment.

It is necessary to understand what the market wants in general, as well as to be able to understand if there are market niche opportunities. For example, two decades ago, wristwatches were considered primarily timepieces that one happens to wear. When microelectronics made it possible for a highly precise watch to be made at a low price, the precision jeweled mechanical Swiss watch industry collapsed. Today, precision and low price are taken for granted and style has become a key parameter. Designer watches for divers that can withstand 50 fathoms of water and look good at the watering hole is one extreme of segmentation that has evolved out of a growing and differentiating wristwatch market. The Swiss watchmakers are back, but are now innovating in style and variety instead of precision.

It is important to understand the market and the customers within the market. Capturing the "voice of the customer" is important, but if the market and the customers in the market cannot be defined, then there is no way of having a clear voice of the customer. One of the problems that a product realization program can encounter early is that there is not a clear definition of the market, which is in the market, and therefore, the voice of the customers in that market.

Thus, one of the key issues for a PRP is to define the nature and identity of the market. The PRP must have the capacity to ensure that the market is defined, segments are articulated, and the customer characteristics within the segments understood.

Product Timing

Time is important for several reasons. It is the denominator of most of the measures that all businesses use to evaluate financial return. Return on investment, internal rate of return, payback period, break-even time, and time-to-profitability are examples of measures that are time dependent. Time is the parameter that determines the length of the cycle of realizing the product. It is the parameter that determines the length of the financial reporting periods and the budgeting periods. For a publicly held corporation, the fiscal "quarter" is the basic unit of time. It is the parameter that determines the period of corporate budget cycles that can control the pacing of product development cycles. It is the parameter that measures the distance between the introduction of products by competitors. It is also important because it measures the changing life cycles of products.

The life cycle is an important metric because it indicates the nature of the dynamics in the industry. For example, the computer workstation industry experiences very short product life cycles. Sun Microsystems, which has been the leader in this market for almost ten years, has a period between new product introductions of approximately six months. A product introduced in January of one year may not only be obsolete by the next January, but more than likely is out of production as well.

If the product realization cycle is longer than the market life of the product, several problems can occur. Consider an example of a product where the lifetime of a product is 12 months and the product realization cycle is greater than 18 months. With these parameters, there can be three or more generations of product in different phases of realization at any one point in time. This poses problems such as: while Product Team A is working on the next-generation product, Team B is working on a product that is to obsolete the product of Team A, but which has yet to be released. This can cause organizational problems (jealousy, competition, etc.). At the same time, it can be an indication of a problem with the product strategy.

One way to solve the short product life cycle/long product realization cycle is to define a product strategy that takes into account the dynamics of the short term while accommodating the natural characteristics of the long term. The concept of product family enters into the product strategy and the approach adopted within a successful product realization process.

Consider the same example described above except that a product family concept has been defined for a period that covers three life cycles. The product strategy represented here is that a product family architecture has been defined that allows there to be three generations of product realized with commonality in key components over this period. There is a team working on the next generation of family that is targeted for three life cyles later. However, the product teams working on the three generations of introduction within the current family can now work together and cooperatively. Sun Microsystems, for example, will introduce new products within a family every six months (where many hardware and software components will remain the same during multiple introductions) but will introduce whole new families of products much less frequently.

As more focus is placed on minimizing the cycle time of product realization, it is important to consider the impact on quality. If a product realization cycle is taking longer than expected or longer than competition will allow, often the reaction is to take shortcuts to get the product into production. Shortcuts tend to result in lower quality, more cost in the factory, and in painful and expensive recall programs once the product is in the field. Time cannot be reduced at the expense of quality.

A PRP is producing products at the right time if:

- The market is ready to receive the product, either because it has been conditioned by previous generations or by the competition.
- The market is surprised by a product with outstanding features that satisfy previously unfulfilled and even unperceived wants.

■ The market's opinion about what constitutes an acceptable product is being influenced by product introductions.

Time is a critical business performance parameter as well because of the importance of the financial results from the product being in the market. One useful measurement parameter is on cash flow because cash is the best indicator of the operational success of a product. Tax strategies and accounting allocations can complicate profitability measures; cash flow is a clear indicator. Either a product is consuming the cash of the corporation or it is contributing to it.

When investment is made in a new product during the realization process, cash flow is negative. After the product is introduced to the market, revenue from the sale of the product becomes a positive cash flow. If the time to payback or the time to positive cash flow is longer than the life cycle of the product, there is a fundamental and strategic problem.

The e-factory requires that the PRP can create the right product for the right market at the right time. The e-factory environment requires a finely tuned product realization business process that includes all of the appropriate resources of the business enterprise and that is tightly coupled with the strategy of the company.

The Customer Relationship Management Process

The customer relationship management process is the process that captures orders for delivered goods and services and that facilitates strong customer loyalty. Customer relationship begins with the contact with a customer and culminates with the receipt of an order. If the customer relationship management is effective, it will result in an ongoing series of orders from customers who are delighted to do business with the company. The activities involved in customer relationship management include:

1. salesforce management
2. customer data management
3. call center management
4. on-line product catalogs
5. product configuration
6. order/proposal configuration
7. order management

Customer relationship management includes the flow of work, material, and information. The flow of information begins with customer contact. The information that must be managed includes order status, production status, process planning, maintenance information, and schedules.

Salesforce Management

Salesforce management refers to the activities associated with directing and supporting the actions of the external and internal sales resources. If a company has an external salesforce working with customers or other channels, there is always a desire to optimize the effectiveness of their time. If a company has an internal salesforce, there is a necessity to have the latest up-to-date information on customers, products, and orders.

As a manufacturing company becomes the e-factory and the channels to market are transformed, eliminated, or directly connected to the factory, the e-factory must be designed to support the salesforce with the proper information. For companies that have a design-to-order type of product or service, e-factory solutions now offer salesforces the ability to have on-line proposal generators, generative product design engines, and mass customization product configurators. This allows an inside or outside salesperson the opportunity to design a solution and create a proposal on-line while with the customers. The salesforce for the e-factory is an extension of the product realization process and the order fulfillment process, rather than just a representative.

Customer Data Management

Customer data management is the creation, control, and exploitation of the database of information about the customers of the e-factory. Typically, a customer database is created because it represents basic billing information about customers, such as the name of key people, addresses, telephone numbers, e-mail addresses, purchase order numbers, and payment status. However, with the advent of data mining software over the last few years, behavioral information about customers, such as the price and frequency of items bought, the category of items purchased, and income category of the customer, has been included in customer databases. This additional information makes it possible to target certain types of marketing messages to the customers in the hope that they will respond with new purchases.

Whatever the content and however it might be used, collecting data about customers will be an opportunity for the e-factory as well as a necessity. The essence of the trends creating the e-factory is to have the enterprise and the entire supply chain focused on satisfying rather specific individual customer needs, and a customer database is essential to that goal.

Call Center Management

A call center is usually a facility where people and technology are working together to answer telephone calls from customers who have questions related to doing business with a company. Generally, call centers are designed to have the majority of the phone calls answered and handled by computer, with only a small percentage of the calls requiring direct human intervention. The same logic that would go into an on-line Internet-based exchange between a customer and a Web site can be used to create an on-line recorded verbal exchange between a caller and the call center.

There are no clear trends indicating whether Internet technology will eliminate the need for call centers. Electronic office technology has not eliminated the use of paper copiers despite years of predictions that it would. However, it is clear that Internet technology can and will augment and amplify the value of call centers. For example, one practice today is for software companies to provide customer service or technical service by phone to customers who have a need to call for help. However, the service representative uses the same on-line Web site for technical and application information that is available to the customer. The result is that the customer is helped by having the service representative either interpret the Web site information or help the customer navigate the Web site. The result is that the customer has been briefly trained in how to use the on-line information, thus possibly reducing the chances that the customer will call back.

On-line Product Catalogs

An on-line product catalog is information such as price, performance, appearance, and fit-for-use features about the products that a company offers to its customers via the Internet. Currently, many companies have basically converted their paper catalogs into electronic catalogs that are available on a commercial Web site. However, the growing trend is to have the catalog become an interactive information exchange that can be tailored to fit the needs of the commercial relationship between seller and customer. For example, an on-line product catalog can be designed so that it will provide a customized list of products, prices, language, and fit-for-use information depending on customer identity. Thus, if a manufacturer forms a corporate purchasing agreement with a large customer which has special pricing, ordering authorization, and delivery instructions, then a customized product catalog Web site can be offered to only that customer with all the appropriate information. This custom Web site might be just a special Web page or instance of the generic Web site, or it might be a separate Web site that is separately hosted. In any case, the on-line product catalog is truly an extension of the e-factory and represents a much more tightly coupled relationship between customer and production than paper catalogs.

Product Configuration

Product configuration refers to the selection of items or features from a product catalog that would result in a working and useful solution for a customer. Typically, a manufacturing company offers a product line with a variety of features and accessories that could be selected in a multitude of combinations, some of which do not work or have no value. One of the challenges for a customer or the salesperson helping the customer is to select a combination that does fit together and that also satisfies the customer's need. Configuring a product solution is becoming one of the more significant tasks for the customer relationship management solution for an e-factory company. How this task is executed in the future will spell the difference between efficiency and success or confusion and failure for the e-factory company.

Order/Proposal Configuration

Configuring an order is similar to configuring a product. The simplest order configuration feature is just having an electronic shopping cart into which customers can "drop" their selected items. However, for customers who are selecting a collection of items that could include hardware, software, service, and intellectual property that are related by function or application, configuring an order becomes more complicated. Issues such as price, interoperability, compatibility, and fit-for-use must be evaluated and approved or rejected as the order is being created. An order configurator picks up where a product configurator leaves off, which is usually in the area of commercial terms and conditions. The order configurator should result in a proposal for a sale to the customer. The customer would then need to make a decision within a certain period of time as to whether he or she wishes to accept or reject the proposal.

An additional feature of an ideal order configurator that should be popular with customers and suppliers alike is the ability to support a negotiation around the proposal. This feature first appeared as on-line auctions for commodities or easily identified products. However, for industries that have more complicated product configurations or service offerings, there is usually more to negotiate than quantity and price. Negotiating on these expanded terms and conditions is the next step in the evolution of having more commercial activity on-line.

Order Management

Once a proposal has been made by a customer and it becomes an order, the e-commerce era has conditioned customers to expect to have information on the status of the order at any time. Although Federal Express was one of the first companies to offer information to its customers about the location and status of their packages at any moment, the world of on-line commerce now expects it of every company and industry. This is a major task because it cannot be done by just having this feature in the Web site. The information about the status of an order must come from throughout the infrastructure of the enterprise. The order fulfillment process must be able to report the status of an order as well as the customer relationship management process. Once the order leaves the premises of the e-factory, its status throughout the logistics system is necessary. Thus, in the e-factory, the status and state of fulfillment of an order for a customer is growing in significance.

Support Processes

Support processes include those activities that are necessary for the successful operation of the primary internal business processes. A support business process begins with a need for services requested by one of the other internal business processes and completes with the delivery of that service. The activities involved in support services include:

1. human resource management
2. financial reporting
3. strategic planning
4. transaction settlement

Support processes include the flow of work and information. The flow of information begins with the business process request for service or from an on-line event that requires processing. The information that must be managed includes financial data, personnel data, legal data, and transaction settlement data.

Human Resource Management

Human resource management is the management of activities such as recruiting, payroll administration, benefits administration, training, counseling, communication, skill set assessment, employment contract negotiation, and performance evaluation. These activities are usually part of every commercial enterprise, whether it manufacturers anything or not. It is a support process because it does not directly involve any process activity that adds value to a product or service. However, it is a critical support activity because if there is a failure in these activities, the other processes tend to grind to a halt.

In the era of the e-factory, there are aspects to the human resource management activities that are new or that are changing dramatically. Skill set assessment, training, and performance evaluation will have to be modified to incorporate the new needs for skills and job definitions. Training requirements are dramatically changing due to the need to incorporate knowledge about new technology tools and new process definitions. In addition, there are new options to outsource payroll and benefits administration from companies that are using Internet technologies to execute and deliver their services.

Financial Reporting

Financial reports are usually required to serve two needs: (1) to communicate to the outside financial world what has been happened to a company financially over a fixed period of time (e.g., a quarter or year), and (2) to communicate to the management of the company what is happening on a much more immediate basis. The Internet in general has made the availability of the external financial reports much more immediate and more widely accessible. The new e-factory technologies are now making key internal financial information much more available to management decision-makers in forms not previously obtainable.

Accounting concepts have evolved to make internal financial information more useful and more readily available. There is a growing trend in the use of activity-based accounting techniques for internal cost accounting purposes. Activity-based accounting principles define how to convert time-period-based functional accounting information into activity-based cost data. The difference here is that cost trends and cost drivers that may be getting out of control can be identified in real-time rather than having to wait for financial reports that are not produced until the reporting period has ended. The activity-based approach allows cost data to be included with the other real-time information being collected by production control systems and other e-factory systems. Now the operational effectiveness of an e-factory can include cost data as well as unit production data and quality data.

Strategic Planning

Strategic planning refers to a broad set of activities that have to do with making decisions about the overall direction of a company in both the short term and the long term. When the concept of strategic planning emerged over the last three decades ago, the typical time horizon for a strategic plan was five years, although some of the longer-term thinkers in Asia often talked about a 25-year or a 50-year plan. In today's world of day traders and the rapid ebb and flow of the stock market, the strategic planning horizon is often measured in quarters — not years. With a shorter time horizon, strategic planning often takes on the look of a tactical plan with the focus being on how fast specific market segments or customer segments can be attacked.

Whatever the appropriate planning time horizon might be, e-factory concepts and capabilities factor into the strategic planning process significantly. Because of the opportunities and challenges of supply chain transformation, a manufacturing company must include in its strategic thinking what the role of e-factory practices, principles, and technologies might be. Employment of any of these e-factory elements will require a major change in company operations and deployment of resources. These are all key strategic issues. One of the goals of this book is to provide a tool that can be used to sort out some of the strategic e-factory issues.

Transaction Settlement

Settling financial transactions is the final step in closing the cycle between customer and supplier. Receiving timely payment from customers for orders shipped and received often spells the difference between financial success and failure. In the world of e-business and the e-factory, with cycle times collapsing in the supply chain, the days of sales outstanding should also be declining. While many companies try to use extended payment terms to partially finance ongoing operations, the era of e-business is rapidly becoming one of cash-and-carry. If customers are expecting to receive immediate order commitment and fast delivery, they will also have to expect to pay in short order. If suppliers are expecting to complete transaction settlement on-line, they will have to offer the appropriate financing discount for immediate payment and likewise charge for the financing expense if an extended payment term is requested.

Irrespective of the pricing or financial term strategy that a manufacturer might pursue, the technology is available for on-line financial settlement. The trend is for more transactions to be settled on-line.

Field Service Business Process

The field service business process is the process that provides a variety of services to customers after they have made a purchase and have had ownership of the products that they have purchased. Field service begins with the schedule or requested service for a customer and concludes with the completion of the service. The activities involved in the field service business process include:

1. spares management
2. human resource management
3. dispatch
4. service planning
5. performance reporting

The field service process includes the flow of work, material, and information. The flow of information begins with customer contact or with scheduled and unscheduled dispatches. The information that must be managed includes spares inventory, staff skill set and availability, field service order status, maintenance information, and resource schedules.

Of all the activities involved in the field service business process, spares management offers the biggest opportunity for improvement using e-factory technologies. The other activities are important and are often in need of change to deal with the impact of Internet technologies. Spares management is the management of the inventory of goods and equipment that are necessary to perform the field service functions of repair and maintenance. For small companies, this might be merged with finished goods inventory management. For larger companies, this might be the largest inventory management challenge in the company. For example, local telephone companies do not manufacture any equipment but they must have hundreds of millions of dollars of equipment spares in their field repair trucks and spares warehouses to ensure that the proper level of service is available.

The challenges for spares management are similar to those of manufacturing inventory management in that the issue is how to match up demand with availability. As the e-factory is bringing build-to-order strategies to many industries, how it can be applied to field service organizations is only now emerging. The best spares management solutions are coming from the warehouse and distribution center management solutions because of the similarities in requirements.

External Business Processes

External business processes refer to those business processes that have extensive contact with and even partial ownership by a supplier or customer. The concept of the external business process is useful because it helps clarify some of the issues that arise as a company thinks about how it should operate in the world of e-business with its e-factory. Because of the dependence of a business process on external influences and external control inputs, the external business process needs to be appropriately designed and operated. The activities typically involved in external business process relationships include:

1. voice of the customer
2. procurement
3. logistics
4. settlement
5. returns and repairs

A brief discussion of these external process interactions follows.

Voice of the Customer

The "voice of the customer" refers to the implied or articulated product or service needs of a customer. In general, this term has been used to describe a key ingredient in a new product development process. The process of capturing the voice of the customer can be viewed as an external process because it requires the involvement of customers in a formal or informal set of activities.

The external process of capturing the voice of the customer has been based on a variety of techniques, from focus groups, to surveys, personal interviews, or demographic simulations. For this external process to be useful, it must be an integral part of the internal new product development process.

With the e-factory, the product and service needs that a particular customer, industry, or entire market can be captured and communicated electronically. The external process can be executed using on-line surveys, interviews, or virtual focus groups such as chat rooms and Web conferences. Because this form of the external process is real-time, on-line, and interactive, it becomes more of a collaboration between the parties requiring a new product and those developing the new product. Thus, in the future, the voice of the customer will become a key part of collaborate product commerce (CPC).

Procurement

While procurement is an important element of the internal order fulfillment process, it inherently involves external activities. The interactions and transactions between suppliers and customers, handled through the procurement process, occur across electronic, physical, and geographic boundaries external to the enterprise.

As the e-factory environment expands, it will create the opportunity to perform not only the electronic execution of external transactions, but will also create the need for electronic negotiation. The area of negotiation is already being addressed in the form of auctions where only prices are negotiated, and all other parameters (e.g., fit-for-use requirements, shipment and scheduling, and other legal issues) are stipulated. However, there are new product services now being brought to market that can support the negotiation of the full range of terms and conditions as well as requirements.

Logistics

Logistics generally refers to the activities of storage and transportation of goods from source of manufacture to customer and intermediate steps. This is a key element of the internal order fulfillment process. Similar to procurement, logistics concerns the involvement of people and organizations external to the enterprise.

While the external portion of the logistics process has traditionally been a key task for manufacturing companies, it is becoming one of the activities that are being dramatically impacted by the advent of the e-factory. There are new companies being created to provide full outsourcing of logistics services so that a manufacturing company does not have to own this process; in effect, the entire logistics process becomes an external process. For those companies that do want to control their logistics process, there are third-party Web-based data services offering a way for smaller manufacturing companies to have the same total electronic parcel tracking capability that companies with world-class logistics capabilities such as Federal Express have developed for themselves.

Settlement

Settlement is the completion of the commercial transaction that begins with a purchase order from the buyer, follows the delivery of the order with an invoice from the seller, and then concludes with the submission of payment by the buyer.

The classical settlement process — and the one that is predominant in the majority of the world — depends on the handling of paper. Paper documents such as purchase orders, shipping notices, invoices, and bank checks have always been the backbone of commercial transactions. With the advent of computers, electronic funds transfer began to replace many financial transfers between financial institutions. With the advent of electronic data interchanges

(EDI), electronic messages began to replace paper purchase orders and invoices. With the advent of corporate credit cards, electronic charges began to replace bank checks as forms of payment. With the advent of e-commerce, the interchange between people in the buying and selling organizations of all these electronic documents can be integrated into a totally electronic set of transactions.

With the advent of the e-factory, the electronic settlement process can be fully automated, thus virtually eliminating the need for human intervention. The e-factory can provide instantaneous notification of shipment, transfer of ownership, receipt, and acceptance. No humans need to be inputting data into a computer to initiate the next step in the transactions.

The settlement process in the e-factory will allow new forms of terms and conditions to be developed and negotiated between parties. Some companies have already begun to use the features of the e-factory to allow electronic settlement to occur for each item produced rather than for each purchase order transaction. Thus, a continuous-flow cellular manufacturing process can have a continuous-flow settlement process.

As the capability to execute the settlement process in real-time permeates entire supply chains, there will be a growing opportunity to use electronic credits and currency analogs as a form of bartering system. There have been proposals of alternative forms of currency (e.g., "Microsoft dollars") that could occur if companies begin trading with each other over a value-added network such as the Microsoft Network (MSN) in monetary units (e.g., Microsoft dollars could only be spent by companies cooperating in the Microsoft network). While this trend has yet to materialize to any degree — due as much to the discouragement by the Federal Reserve Board as to the lack of interest by potential customers — it has created renewed interest in exchanges dealing in barter of goods and services.

Returns and Repairs

Returns and repairs refer to the return by customers of the goods that they received in fulfillment of their orders. The causes for these returns usually have to do with customers not being satisfied with what they received, either because their perceptions of what they were buying were different from reality, they changed their mind about what they wanted to buy, or the product failed to perform according to specification. There is nothing new about the nature of returns in the e-business era other than the fact that returns rates are extending to be significantly higher for e-business transactions than for retail store transactions. Either expectations are being inappropriately set because insufficient information is being transmitted to the customer via the Web site or buying behavior on the Web is more impulsive than normal retail commerce.

The e-factory thus needs to be able to deal with returns and repairs — especially in light of the higher rates of return. Thus, the solutions for order fulfillment and customer relationship management cannot support just one-way directional flow, but must also support two-way directional flow.

Summary

For the e-factory to be a success, the newly available synergy between process and technology must be achieved. The structure of a business process determines the limits to the degree of efficiency that can be achieved, whether or not technology is used. However, technology can accelerate the execution of business processes and reduce cycle times. As the capabilities of technology continue to expand, there is increased pressure to restructure and redesign business processes to ensure that they can benefit from the synergy with technology. Process structure can impede or facilitate receiving the benefit of technology. Old or existing processes may need to be reengineered. New business processes must be designed to take advantage of the benefits that technology can offer.

One benefit is real-time information sharing between legacy systems and new packages, between different companies, and between different data types and structures. Another benefit is in improved workflow management and workflow control. Business processes, both internal and external, can be structured to take advantage of the promise that technology offers to facilitate business transactions, to facilitate process planning and scheduling, and to collaborate with business partners. All of the key business processes in the e-factory will have to be transformed or changed dramatically to take advantage of new technology and to improve their performance to be able to respond to the demands of the Internet environment. The e-factory requires both process innovation and technology innovation. The process should be designed with the technology solution included — not with technology added as an afterthought.

Chapter 5

Fission and Fusion in the e-Factory

Purpose of This Chapter

There are several forces that are conspiring today to create massive change in the allocation of where value is added in the supply chain and in the factory environment. These forces are created by economic trends, technological advances, global competitive pressures, and new strategic thinking. These forces are causing transformations that on one hand are causing functions, business processes, corporate structures, and even industries to disassemble (i.e., the fission part of the formula); and on the other hand, these forces are also causing transformations that see new types of integration or fusion of business processes, corporate entities, and industries. The purpose of this chapter is to examine these trends and provide some insight into the impact that these counter-balancing trends are having on the e-factory. These trends should be an important consideration during the creation of the e-factory roadmap.

Current Trends

The diagram in Exhibit 5.1 is a simple model of what an order fulfillment process might look like in a manufacturing enterprise that has a traditional structure of production, product design, purchasing, and logistics functions. This simple setup can be used to illustrate the type of fission and fusion occurring in various industries. It is also illustrative of the type of thinking that should be going on when an enterprise is developing its manufacturing strategy.

A key element in the e-factory roadmapping methodology (described in Chapter 8) is the development of operational scenarios. Each scenario is a description of the activities and the human, machine, and data resources

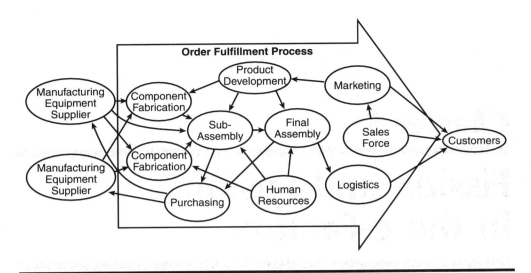

Exhibit 5.1   Generic Integrated Enterprise Example

involved in each activity that are required to execute the mission for each business process. As these scenarios are created, a key question that must be asked is how much fission or fusion should there be in the new e-factory.

There are a variety of forcing factors — some of which have been in place for quite a while and some which have just appeared — that are causing fission and fusion in the supply chain.

The older factors have been driven by companies seeking ways to improve the quality and productivity of their business processes and their supply chains. These include:

- top-tier assembly
- multi-plant scheduling
- demand communication and planning

Top-tier assembly is the concept of a manufacturing enterprise organizing its suppliers into tiers where the top tier is performing key sub-assembly activities that integrate the materials supplied by second- and third-tier suppliers. This reduces the number of suppliers that the manufacturing enterprise has to work with and streamlines the planning process. However, it also creates the need for new types of activities such as multi-plant scheduling.

Multi-plant scheduling is part of a supply chain management concept of demand synchronization. The basic premise is that inventory throughout the supply chain can be minimized by managing all the plants in the supply chain as if they were just cells in a production line. The strategy is to synchronize the flow of material through the supply chain plants just as it was flowing through a continuous-flow, just-in-time factory. This is a challenge for multi-plant companies — much less getting multiple companies in a supply chain to collaborate on scheduling their plants.

While multi-plant scheduling is a high-level management challenge, a lower level challenge is real-time order management. With the advent of e-business creating an expectation in all customers that a supplier should be able to track and report the status of each individual order, comes the need to provide that information throughout the supply chain. And as more manufacturing companies move to a build-to-order capability, providing real-time order status information is becoming a necessity.

The more recent factors have been driven by companies seeking to exploit the potential of the Internet, to be more responsive, to market needs, and to extend the reach of an enterprise outside the normal boundaries. These factors include:

- real-time order management
- electronic purchasing
- channels transformation

Real-time order management responds to the need in the world of e-business for the status of the commercial transaction between two parties to be available at all times. This need began with the experience that people have had with the very successful on-line retail operations of businesses such as Amazon.com and Dell.com. As business-to-business transactions became more prolific on the Internet, the expectation that the status of an order between two commercial organizations should be available at all times also became an expectation.

Electronic purchasing is becoming an important factor for a variety of reasons. This is much more than just being able to implement EDI via the Internet. Included in this factor is having suppliers' product catalogs on-line, having the customer's contract terms and conditions on-line, having the ability to purchase on-line, and having the ability to negotiate terms and conditions on-line.

Channel transformation is another factor that significantly affects whether an enterprise is facing fission or fusion. With all the money invested in electronic retailing, purchasing exchanges, and Web-front stores, the role of traditional selling channels such as retail chains and distributors is changing. While the predictions of these channels disappearing totally are probably far-fetched, there is clearly a transformation going on in their purpose and structure. As these transformations are occurring, there are opportunities and challenges for the manufacturing companies that have traditionally sold through them.

Fission in the Enterprise

Fission in the enterprise is being caused when companies focus on core competencies and decide that they should not own processes that do not add significant value in their supply chain.

The term "fission" is being borrowed from physics because it conjures up the image of one corporate entity being blown apart because of internal pressures or outside influences to form two or more corporate entities that did not previously exist. While there are probably no laws of nature at work here, there is a phenomenon that is clear and growing. It really started in the early 1990s with a trendy concept called "outsourcing." The concept then was primarily aimed at the reducing the cost of corporate information technology organizations.

The outsourcing concept then was to take all of the people and capital associated with the information technology organization and sell it to another enterprise that specialized in managing such organizations. The cost-saving was supposed to come from the fact that the specialty organization would be able to run the information technology group more efficiently. The working rule-of-thumb was that there should be a cost reduction of 20 percent to the enterprise that sent its process to an outsourcing company. The outsource company would achieve the cost reduction, usually by taking the existing IT group and reengineer all the business processes, reduce the cost of labor through replacement or attrition, and focus on providing only that work which was explicitly specified in the outsourcing contract. If any new work was to be purchased by the outsourcing company, then a new contract with new pricing would have to be negotiated.

Needless to say, that trend waned after a while because the cost benefits for the outsourcing enterprise often failed to materialize as the cost of any new work tended to offset much of the savings from the services originally outsourced. However, the concept of outsourcing did take hold in other areas, primarily because it created the opportunity for an enterprise to specialize in key core competencies while buying other services that were needed by the corporation to function.

An illustration of enterprise fission is presented in Exhibit 5.2 by beginning with the model from Exhibit 5.1. Several examples of fission are suggested in Exhibit 5.2. The production process has been reduced to final assembly and test with all component and subassembly manufacturing outsourced to other companies. All shipping and inventory management has been outsourced to a logistics enterprise. The sales functions have been outsourced to a sales and distribution enterprise. Marketing activities have been outsourced to a marketing and advertising enterprise. Product development has been outsourced to an engineering services enterprise that specializes in new product development. The human resources support activity has been outsourced to a human resources service enterprise. The enterprise that now remains has as its core competency the final assembly and test of a finished product. This example may appear to be extreme, but it is not too different from that now being seen in the semiconductor industry with companies that are known as silicon foundries.

In the following, some of the key issues and opportunities related to the e-factory roadmapping due to outsourcing of different functions within key business processes are described.

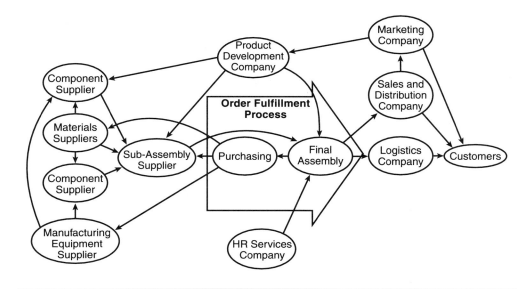

Exhibit 5.2   Enterprise Fission

Outsourcing Manufacturing

Manufacturing outsourcing has been popular for years. IBM was one of the first companies to resort to outsourcing of manufacturing when it decided to launch its now-famous personal computer with contract manufacturing 20 years ago. The latest glowing success story is Cisco Systems, which outsources almost all of its manufacturing except for some final test and assembly.

There are several reasons why an enterprise whose mission is to deliver goods and services to a customer would resort to outsourcing of manufacturing. One reason would be that the enterprise wishes to focus on marketing, product development, or service as core competencies rather than manufacturing. Another is that product technology might be changing so fast that the enterprise may not want to risk investing in a manufacturing base that might become obsolete very quickly, but would rather take the risk that there would be an available base of manufacturing companies that could provide the service needed. The assumption that there would be suppliers available that can provide the manufacturing on an outsource basis is based on the assumption that there would be enough of a market for outsourcing services that there would be risk capital seeking the companies that wished to serve the market.

When an enterprise decides to outsource manufacturing, e-factory concepts apply to both the enterprise buying the manufacturing services and the enterprise providing them. Of course, the enterprise providing manufacturing services is amenable to e-factory concepts because it owns the physical factory. The enterprise buying the outsourcing services may be in the situation of having to manage a virtual e-factory if it wishes to have certain kinds of control over its new nodal supply chain network.

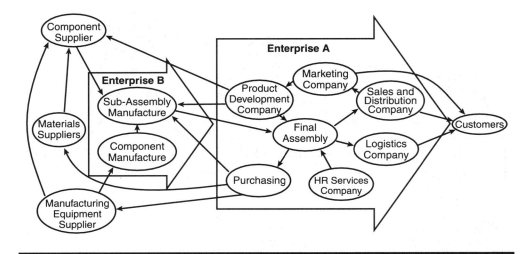

Exhibit 5.3   Outsourcing as Enterprise Fission

Using the example in Exhibit 5.3, if Enterprise A decides to outsource the manufacturing of a new product to Enterprise B, Enterprise A will have to decide who has authority over production process improvements, inventory levels, order management, and production scheduling. If Enterprise B controls these factors, then Enterprise A has very little to be concerned with other than giving Enterprise B a list of which products it wants manufactured and a time when they should be delivered. If Enterprise A controls these factors, then Enterprise A becomes supply chain focused, with Enterprise B becoming a manufacturing node in Enterprise A's tightly coupled web. In this case, Enterprise A has to manage many of the e-factory issues for the node that is Enterprise B. Enterprise A must operate as if it had a virtual e-factory. Flowing from this situation is the need for Enterprise A to have an e-factory capability and therefore an e-factory roadmap.

To determine what degree of e-factory fission is required, several key questions must be answered by the enterprise buying the outsourced manu-facturing services. These key questions include:

- Who will do purchasing?
- Who will do production scheduling?
- Who will manage logistics?
- Who will do supplier selection?
- Who will assure quality?
- Who will perform invoicing?
- Who will manage order status?

The answer to these questions shape the nature of the e-factory require-ments for both Enterprise A and B. Therefore, there is a need for an e-factory roadmap for both Enterprise A and B. The roadmap for Enterprise A should be developed in conjunction with Enterprise B. If Enterprise B already has certain e-factory capabilities in place, then Enterprise A can more easily decide how much of the e-factory it needs to implement. If Enterprise B does not have an e-factory roadmap, then roadmap development should be done in conjunction with Enterprise A.

The answer to these questions will also determine the special e-factory information technology requirements. For example, with respect to the issue of who will do purchasing, if Enterprise A is a very large enterprise and Enterprise B is a smaller enterprise, Enterprise A may get a better price discount on purchased material than B. Also, Enterprise A may want to control purchasing to ensure that availability of material is high or that inventory at B remains under control. A summary of the potential impact on the e-factory requirements is given in Exhibit 5.4.

Outsourcing Maintenance

Outsourcing maintenance has only recently begun to emerge as a viable option for manufacturing companies. Several companies that make capital equipment are beginning to offer this type of service as part of their total service offering. Other companies already in the service industry are beginning to offer these services independently.

There are several reasons why an enterprise whose mission is to deliver goods and services to a customer would resort to outsourcing of maintenance. One reason would be that the enterprise wishes to focus on manufacturing and not have to have the staff or spares inventory on hand waiting for a repair problem. Another reason is that manufacturing technology might be changing so fast that the enterprise may not want to risk investing in spare inventory, maintenance training, or substantial equipment downtime that would result from internal maintenance.

When an enterprise decides to outsource maintenance, e-factory concepts apply to both the enterprise buying the maintenance services and the enterprise providing them. Of course, the enterprise providing maintenance services is amenable to e-factory concepts because it owns the physical inventory and human resources. The enterprise buying the outsourcing services may be in the situation of having to manage a virtual e-factory if it wishes to have certain kinds of control over its extended enterprise that is now part of its supply chain.

For example, if Enterprise A decides to outsource the maintenance of a new production line to Enterprise B, Enterprise A must decide who has authority over spare ownership, repair process improvements, spare inventory levels, resource deployment, and maintenance scheduling. If Enterprise B controls these factors, then Enterprise A has very little to be concerned with other than giving Enterprise B a list of which equipment and systems it wants maintained and the minimum acceptable response time. If Enterprise A controls these factors, then Enterprise A becomes supply chain focused, with Enterprise B becoming a maintenance service node in Enterprise A's tightly coupled web. In this case, Enterprise A has to manage many of the e-factory issues for the node that is Enterprise B. And Enterprise A has to operate as

Exhibit 5.4   Manufacturing Outsourcing Impact Table

Activity	Lead	*Enterprise A e-Factory Requirements*	*Enterprise B e-Factory Requirements*
Scheduling	A	Generate schedule, receive production capacity and availability information from Enterprise B	Receive schedule information electronically, execute schedule, report schedule status, report production availability
Logistics	A	Negotiate and administer contracts with transportation companies, generate and transmit shipping requirements	Receive shipping plans, execute plans, generate shipping bills, report shipping results
Supplier selection	A	Negotiate and administer contracts with suppliers, generate and transmit material delivery requirements	Receive material plans, execute plans, generate purchase orders, report quality results
Quality	A	Develop quality strategy and perform quality audits and certifications	Execute quality strategy and
Invoicing	A	Develop invoicing terms and conditions, create invoices and submit to customers	Submit invoice information to Enterprise A
Order status	A	Collect data, create status database, provide real-time access to database	Collect and transmit data to Enterprise A database
Plant scheduling	B	Receive schedule status, transmit schedule inputs	Develop schedule, execute schedule, report schedule status
Logistics	B	Receive shipping reports, transmit shipping inputs	Negotiate and administer contracts with transportation companies, generate shipping plans, execute plans, generate shipping bills, report shipping results
Supplier selection	B	Receive material plans, execute plans, generate purchase orders, report quality results	Negotiate and administer contracts with suppliers, generate and transmit material delivery requirements
Quality	B	Execute quality strategy and	Develop quality strategy and perform quality audits and certifications

Exhibit 5.4 Manufacturing Outsourcing Impact Table (continued)

Activity	Lead	Enterprise A e-Factory Requirements	Enterprise B e-Factory Requirements
Invoicing	B	Execute quality strategy, submit invoice information to Enterprise B	Develop invoicing terms and conditions, create invoices and submit to customers
Order status	B	Collect and transmit data to Enterprise A database	Collect data, create status database, provide real-time access to database

if it had a virtual e-factory. Flowing from this situation is the need for Enterprise A to have an e-factory capability and therefore an e-factory roadmap.

To determine what degree of virtual e-factory is required, several key questions must be answered by the enterprise buying the outsourced manufacturing services. These key questions include:

- Who will schedule maintenance?
- Who will hold spares inventory?
- Who will collect data and how?
- What level of uptime is to be guaranteed?

The answers to these questions shape the nature of the e-factory requirements for both Enterprise A and B. Therefore, there is a need for an e-factory roadmap for both Enterprise A and B. The roadmap for Enterprise A should be developed in conjunction with Enterprise B. If Enterprise B already has certain e-factory capabilities in place, then Enterprise A can more easily decide how much of the e-factory it needs to implement. If Enterprise B does not already have an e-factory roadmap, then this should be done in conjunction with Enterprise A.

Outsourcing Logistics

Logistics outsourcing has been one of the fastest growing new areas of outsourcing during the last five years. A variety of transportation companies, overnight delivery services, and new start-up companies are now offering a broad range of logistics planning and execution services to the general manufacturing marketplace.

There are several reasons why an enterprise with a mission is to deliver goods and services to a customer would resort to outsourcing of logistics. One reason would be that the enterprise wishes to focus on manufacturing and not invest in the staff or transportation resources to serve its transportation needs. To achieve the level of logistics service required by the Internet economy, the necessary level of investment in logistics resources might show a low level of return due to under-utilization of the resources.

When an enterprise decides to outsource logistics, e-factory concepts apply to both the enterprise buying the logistics services and the enterprise providing them. Of course, the enterprise providing logistics services is amenable to e-factory concepts because it owns the logistics assets. The enterprise buying the logistics services could be in the situation of having to manage a virtual e-factory if it wishes to have certain kinds of control over its supply chain.

For example, if Enterprise A decides to buy logistics services from Enterprise B, then Enterprise A must decide who has authority over transportation services, scheduling of shipments, tracking of shipments, packaging, intermediate storage or warehousing, and delivery confirmation. If Enterprise B controls these factors, then Enterprise A can be relieved of most logistics decisions and can focus on other processes. If Enterprise A controls these factors, then Enterprise A becomes supply chain focused, with Enterprise B becoming a logistics service node in Enterprise A's tightly coupled web. In this case, Enterprise A must manage many of the e-factory issues for the Enterprise B node. Enterprise A has to operate as if it had a virtual e-factory, which included logistics. Flowing from this situation is the need for Enterprise A to have an e-factory capability and therefore an e-factory roadmap. Some logistics companies are offering services that can provide a virtual e-factory capability to an Enterprise A.

To determine what degree of virtual e-factory is required, several key questions must be answered by the enterprise buying the outsourced manufacturing services. These key questions include:

- Who will schedule and plan logistics?
- Who will hold and distribute inventory?
- Who will collect data and how?
- What level of performance is to be guaranteed?

The answers to these questions shape the nature of the e-factory requirements for both Enterprise A and B. Therefore, there is a need for an e-factory roadmap for both Enterprise A and B. The roadmap for Enterprise A should be developed in conjunction with Enterprise B. If Enterprise B already has certain e-factory capabilities in place, then Enterprise A can more easily decide how much of the e-factory it needs to implement. If Enterprise B does not already have an e-factory roadmap, then both roadmaps should be done collaboratively.

Outsourcing Product Realization

The outsourcing of product realization and development has been occurring only in selected industries. Most manufacturing companies view product realization as a necessary core competency. However, there has been fission of the product realization process in several industries, such as pharmaceuticals, consumer goods, and Internet electronics. There are a variety of reasons for this type of enterprise fission.

One reason is that the speed of change in product technology or in the voice of the customer is so fast that the business process of product realization has to be completely unique and independent of the order fulfillment process. Many consumer goods manufacturing companies outsource product development for just this reason.

Another reason is that the investment in new product development is so expensive that there has to be some consolidation of investment or collaboration in approach between manufacturing companies, and the easiest way to do this is through a third party dedicated to product realization. The pharmaceutical and semiconductor industries are beginning to invest in companies that only do product development.

A third reason is that some product realization companies have invested in rare skill sets that are not generally available to manufacturing companies. Thus, outsourcing of portions of the product realization process can complement the skill sets of the manufacturing company.

When an enterprise decides to outsource product development, e-factory concepts become critical to the operation of the relationship. Having the ability for rapid, electronic, and collaborative communications between the supplier of the service and the customer of the service is essential. In fact, the relationship between the two parties has to be more of a partnership. To determine what degree of virtual e-factory is required between the two companies, several key questions must be answered. These key questions include:

- Who will own the intellectual property being developed?
- Who makes the intermediate and final decisions about design approach and solution?
- How will the design supplier be compensated?
- Who is responsible for the voice of the customer?

If part or all of product realization is to be part of an outsourcing arrangement, an e-factory roadmap should be developed in collaboration with the outsourcing partner.

Outsourcing Testing

Testing services have been available for many years in many industries. However, there may be an even greater amount of testing service needed in the future because of the emergence of e-factory requirements and because of the growing requirements for tests that are being created by regulatory and legislative agencies around the world.

Testing often requires specialized equipment, specially trained professionals, and special facilities. For some industries, the investment in test resources is prohibitive except for the very largest companies or those specializing in testing as a core competency and as a service offering. For other industries, testing is an essential part of process control and must be a core competency.

When an enterprise decides to outsource testing, e-factory concepts apply to both the enterprise buying the testing services and the enterprise providing them. Of course, the enterprise supplying testing services is amenable to e-factory concepts because it owns the testing resources and produces the data and knowledge required by the customer. An e-factory roadmap is required for both companies because of the importance and timeliness of the information being generated. The same business concepts being developed by the maintenance services and logistics services companies that use the Internet as a facilitating medium are also being developed by the testing services companies. Vendor-managed inventory can now become vendor-managed testing and vendor-managed logistics.

To determine what degree of virtual e-factory is required after this level of fission, several key questions must be answered by the enterprise buying the outsourced testing services. These key questions include:

- Who will schedule testing and what is the turnaround time?
- Who will own the test data?
- How will the data be transmitted, managed, and preserved?
- Who defines the tests?

If part or all of the test functions in a company are to be part of an outsourcing arrangement, an e-factory roadmap should be developed in collaboration with the outsourcing partner

Outsourcing Customer Relationship Management

The outsourcing of customer relationship management is a key strategic question that is becoming more important as the Internet economy continues to evolve. While the outsourcing of other processes and functions discussed herein can often be treated simply as a strategic make or buy decision, customer relationship management is becoming more important as a core competency for companies that never had it before as a key business process.

Prior to e-business, the relationship between a manufacturing company and its customers was a function of the structure of its supply chain. If the company sold through distribution channels, it was not in direct contact with end users and only had to maintain a relationship with its distributor partners. If the company sold directly to other companies that may not have been the final end users either, then there was a field salesforce that maintained customer relationships. The new market rules of e-business are now making business-to-business and business-to-consumer relationships a strategic requirement for manufacturing companies. As manufacturing companies begin to evolve to the e-factory and e-supply chain, how to create and execute customer relationship management becomes a critical strategic decision. For many if not most companies, customer relationship management should be a key element of the e-factory. For others, there may be a need to outsource the process.

The entire field of customer relationship management is relatively new. The term itself became popular in the late 1990s as an industry name for the various software applications that were being brought to market to support Web-enabled sales force automation, call centers, and data mining applications. Since then, it has grown to include all the functions and software tools that can be used to provide on-line customer interactions and marketing.

Given all the trends contributing to the emergence of the e-factory, now every manufacturing enterprise and every company operating in a supply chain must consider how much customer relationship management is required and how can it be implemented. This new strategic need contributes as much to fusion in the supply chain as it does to fission in the enterprise. Because it is a new function for so many manufacturing companies, fission or outsourcing may not be the correct term. For many companies where this function does not exist, a contract for this service will have to be established if the capability cannot be built internally fast enough. Thus, outsourcing or buying a customer relationship management service is now a strategic issue for many companies planning an e-factory strategy.

The key questions that need to be answered as a company considers buying or outsourcing customer relationship management include:

- Who will own and maintain the customer database?
- What are the necessary links between the service and the rest of the e-factory?
- Will all contacts with customers and channels be handled by the service?
- What are the performance parameters that will be used to measure the success of the service?

As the roadmap for the e-factory is developed, the role and source of the customer relationship management functionality should be explicitly included. The roadmap itself could be used to assist in deciding whether to build the capability internally or use an outside service.

Outsourcing Information Technology

The concept of outsourcing has been around for decades and has undergone periodic changes in form and content. The concept of outsourcing information technology launched a new industry in the early 1990s. The concept then was that one company, usually a very large company, would sell all its information resources (including people, facilities, and technology) to a third party. The third party would then sell information services back to the company at a price that was lower (usually 20 percent lower) than the old internal cost of the same services. This same basic concept continues today for information services as well as Internet and Web site services. The size of the information technology outsourcing market has grown steadily over the last decade.

The problem that information technology outsourcing services have had in the past is that the concept depends on a certain level of stability in the need for information technology by the company being served. Of course, information technology is not usually thought of as a stable phenomenon and as the need for new services were surfaced by the served company, the outsourcing company would have to charge competitive rates to implement new capabilities. As a result, many outsourcing projects did not achieve the kind of long-term benefit that was originally forecast.

Several key questions must be answered by the company buying the outsourced information technology services. These key questions include:

- Who will own the information technology resources?
- Who will pay for changes in the information technology resources?
- How will new information technology requirements be identified, communicated, purchased, and received?
- What level of service is to be guaranteed?

There are a growing number of options for information technology outsourcing from which to choose. The e-factory roadmap can be used to assist in evaluating business process-driven options and in the final decision-making process.

Outsourcing Field Service

The outsourcing of field service is in many ways similar to the outsourcing of logistics or maintenance. For many companies, field service is not a strategic core competency but more of a business function that is a necessary part of the cost of doing business. For companies that view field service as not only an essential ingredient in satisfying customers, but also a primary goal, outsourcing it may not be strategically wise.

The fundamental issue is what role field service plays in a corporate strategy. There are many companies whose only business is field service and they offer this capability on an outsource basis to those companies that do not want to own it. It all comes down to deciding where a company can and wants to add value in the supply chain.

The key questions that must be answered before buying outsourced field services include:

- Who will set policy, terms, and conditions for field service?
- Who will set prices?
- How will the outsourced field service communicate and operate with the e-factory?
- Who will determine what services will be offered by an outsourced field service?
- What will be the relationship between the field service organization and competitors?

Independent or outsourced field service organizations are frequently found in many industries. However, in the new era of the dynamic supply web, there will be new opportunities for rethinking whether this is an area where a company should or should not add value.

Outsourcing Human Resources

Outsourcing human resource management activities has been an active area for many years in the general field of outsourcing. It has accelerated with the introduction of Internet technologies because so many people now have access to the Internet through personal or company-owned computers. The Internet offers the dual benefits of easy access to a corporate knowledge base while still maintaining privacy and, to some degree, anonymity.

The key questions that must be answered by the company considering buying outsourced human resource management services include:

- Who will own the human resource knowledge base?
- Who will manage and maintain the human resource knowledge base?
- How will individuals access and use services?
- How will the company ensure privacy and fail-safe operation?

The e-factory will depend on a responsive and flexible human resource business process because the speed with which it responds to human resource issues must match the speed required to respond to the marketplace.

Fusion in the Supply Chain

One of the fundamental strategic questions that a company must answer for itself periodically is: how can it add more value to the products and services that it delivers to its customers? The purpose of adding more value is, of course, to earn a larger portion of the profit available in a supply chain. The simple answer to the question is to integrate either forward or backward in the supply chain, to replace a supplier, or to sell to the next-tier customer. This financially driven supply chain "fusion" has been the stimulus for mergers and acquisitions for years in a variety of industries.

However, the advent of e-business and the Internet has created a new stimulus that is technology driven but which creates new opportunities for any company in any link in a supply chain to have direct access to final users of their products. Now that any company has the ability to be on-line with any customer from any forward link in the supply chain, a new strategic question arises: how to become more customer centric and customer focused using the Internet. And the fact that a company can be directly connected on-line to any customer creates new opportunities for "fusion" in the customer facing, forward elements of the supply chain, the "channels" to the market.

In fact, a special term, disintermediation, has been coined to describe this channel fusion.

Disintermediation is the elimination of the intermediaries between a company and the final user. In traditional industrial product industries, intermediaries are the sales and distribution channels that provide the coordination, selection, communication, and exchange functions with end users. End users typically purchase through local, independent sales representatives, who provide technical knowledge, applications assistance, and supplier selection. For customers with high-volume, repeat purchases, distributors that have local warehouses provide fast delivery. The sales representatives and the distributors are intermediaries that represent additional costs in the supply chain in exchange for the local access to knowledge and inventory.

However, if the supplier of the product can provide real-time access and can fill orders quickly, the value of the sales representative or distributor diminishes. The real-time access can be provided by Internet technology. The fast filling of orders can only be provided by either on-the-shelf inventory or fast order fulfillment processes. That is why process and technology have to combine for this type of fusion to occur. Examples of fusion include on-line personal computer retailing, on-line ticket sales, on-line car sales, and on-line office supplies.

This trend of fusion does not clearly reveal what the final outcome or best practices might be. In fact, the trend could be described as the extending of the enterprise forward into the supply chain using customer relationship management technology while at the same time maintaining and supporting the traditional intermediary-based channels to market.

Another way to look at fusion in supply chain management is to look at the blurring and intermixing of information, workflow, and decision-making that is taking place between companies that operate in a supply chain. The term "fusion" is being borrowed from physics because it conjures up the image of resources from several different companies being smashed together to form an entirely new resource, similar to the creation of helium from the fusing of hydrogen atoms. Interestingly enough, what may be one company's fission might be another company's opportunity for fusion. This obviously calls into play the cosmic relevance of the e-factory.

Supply chain fusion is having an impact in several areas of an enterprise's structure. The diagram in Exhibit 5.5 illustrates one example of fusion in a supply chain. In this example, an order fulfillment process has been created by the fusion of design, purchasing, planning, marketing, and sales processes, while all the manufacturing and human resource management processes are external to the process via outsourcing. This factory-less order fulfillment process is an independent company that has decided to add value by organizing and managing supply chains. In many ways, this is the Cisco Systems model because most of their products are manufactured by suppliers with whom they have established special external business processes for managing the flow of information and material.

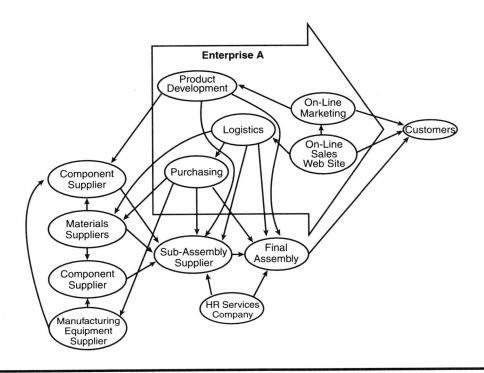

Exhibit 5.5   Supply Chain Fusion

One fusion phenomenon is the sharing of information between companies within a supply chain and within each enterprise. The sharing is occurring between legacy information systems and the new enterprise systems that many companies have been implementing over the last few years. There is also a need to share information between plant scheduling systems, design and manufacturing systems, order management systems, warehouse systems, product development systems, and point of sale systems. This sharing must be in real-time for it to be effective, and it must be between a variety of data types and structures.

Another area where fusion is occurring is in workflow management. As companies learn how to share information between each other to support the coordination of demand and supply in the supply chain, they will also learn how to manage the flow of work between themselves. With the sharing of data for the purpose of coordinating actions and decisions that need to be made for the sake of the supply chain, there comes the need to manage the flow of work between companies. This is possibly a subtle type of fusion, but it is real. If Enterprise A is asking its supplier, Enterprise B, to manage the inventory that Enterprise A is purchasing from Enterprise B on Enterprise A's premises (otherwise known as VMI or vendor-managed inventory), then Enterprises A and B have decided to jointly manage the workflow related to inventory management.

A third area where fusion is occurring is in decision-making, especially for areas such as production scheduling, product development, inventory management, and quality assurance. Decision-making fusion is occurring in the form of new contract terms and conditions that are negotiated between supplier and customer and in the form of the information sharing being done with new supply chain management and enterprise software.

Two kinds of fusion in the supply chain are emerging. There is a virtual fusion where business processes from different companies in a supply chain "fuse" for short periods of time to collaborate on a joint project or to experiment with a new business or technology concept. There is physical fusion where two companies merge their operations or they jointly form a third company that will be responsible for the execution of a fused set of business processes. The formation of dot.com companies by one or more large corporations is an example of physical fusion.

Product Development Fusion

Product development fusion refers to companies working together, either in supply chain teams or in newly created design companies, to create new products that can be manufactured by one or more of the founding companies.

The collaborative teaming approach happens frequently between suppliers and customers in a supply chain when there is opportunity for mutual benefit in either cost, cycle time, or quality. There are new software tools on the market that are the successors to workgroup management and product data management software that allow product realization teams from different sites and different companies to collaborate on new projects.

New venture capital-backed companies are being formed in industries where there is a need for innovation in design or investment risk management. These include the pharmaceutical, semiconductor, telecommunications, and petrochemical industries.

Purchasing Fusion

Purchasing fusion refers to the consolidation of purchasing functions for multiple companies in either an extended purchasing process of one company or in the creation of a virtual integrated purchasing group. Technology is now available to make it possible for one company to manage and exercise the buying power of a collection of companies that by themselves would not have as much influence with suppliers. The same technology is making it possible for a group of companies to operate as one, in a virtual sense as a buying cooperative or as a purchasing consortium. This trend is also fast emerging in a variety of forms. One of the fastest growing is in the form of an e-market. The role and growth of e-markets are discussed in Chapter 11.

Channel Fusion

Channel fusion refers to the transformation of the marketing, selling, and distribution channels to market. This involves the integration of these functions within different companies in ways that are new or different from the past. One type of channel fusion is the extended enterprise. Wherein, manufacturing companies are selling directly to end users over the Internet, service companies are selling directly to end users over the Internet, and other companies are getting new product ideas over the Internet. Channel fusion also refers to single product or industry channels expanding to include other products or services such as the businesses of Amazon.com and Dell.com. These retail Web sites are selling a variety of consumer products that extend beyond what the Web site originally was created to do. Channel fusion also refers to the nature of the interface between the manufacturing company and the channels. For example, an e-factory in some industries will have to have an Internet-based interface with distributors, with final end users, with virtual distributors and e-tailers, and with special customers. The degree of commonality of these different interfaces depends on the features and functions that these different channels require. There will likely be a significant amount of commonality of functionality required.

Implications of Fission and Fusion on the e-Factory Roadmap

When building the roadmap for the e-factory, it is essential that it reflect whether there is fission and/or fusion as part of the strategy. A strategy of enterprise fission will require that the roadmap include the involvement and/or consideration of the external partners and the business processes that they represent. A strategy of supply chain fusion will require that the business processes to be integrated into the enterprise will be included in the development of the roadmap.

One of the conclusions that can be drawn by considering the impact of enterprise fission and supply chain fusion is that the role of the e-factory as either a controlling node or a responding node in the supply chain must be clearly defined. The diagram in Exhibit 5.6 illustrates how an e-factory node, even without any production assets, would have to be in a controlling position. As fission and fusion continue to cause transformations in the structure of supply chains in different industries, the supply chain node that produces goods and services is a critical element in the critical path. The eventual structure of the supply chain is one of the key factors in determining the requirements that drive the e-factory node.

Among other requirements, the e-factory must communicate with other nodes in the supply chain, with new entities that are forming, and with new

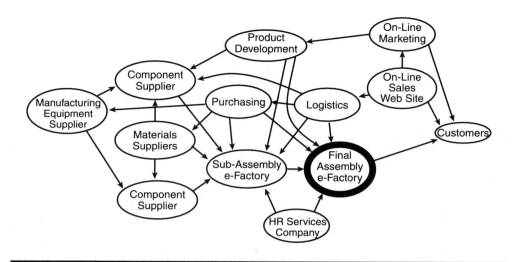

Exhibit 5.6   e-Factory as Controlling Node

channels to market. The e-factory will likely have to communicate directly with the final end user of its products, as well as with any other channels to the end user. The e-factory node must control internal processes, external processes, and the interfaces between internal and external.

Summary

In many ways, supply chain fission and fusion are occurring simultaneously for almost the same reasons. As management teams look for ways to improve their profitability and their return on investment, they are also looking for ways to increase the amount of value they add in the supply chain. For some, this means enterprise fission or outsourcing functions and processes that offer no strategic advantage. For others, this means supply chain fusion or integrating forward and/or backward into their supply chain.

Fission and fusion are creating new requirements for the e-factory. When enterprise fission occurs, there will be requirements for the e-factory to control the external processes as well as the internal processes. Enterprise fission might also create the need for a virtual e-factory to exist that consists of one or mode supply chain nodes. The virtual e-factory must control external processes and interact with other e-factory nodes in the supply chain. For the company that is planning supply chain fusion, the e-factory requirements must include processes and activities that do not yet exist within the enterprise. The e-factory roadmap approach must capture all of these new internal and external requirements.

Chapter 6

e-Factory Technology Solution Sets

Purpose of This Chapter

As technology and process trends accelerate the introduction of new technologies into the e-factory, decision-making about which software and hardware suppliers to use becomes more complex. Over the last half of the last decade there was a raging debate associated with information technology strategies as to whether a company should use a single software provider for all its business application and control needs, or whether there should be a "best-of-breed" approach. The rapidly growing enterprise resource planning (ERP) companies such as SAP and Oracle were of course advocating the single supplier enterprise approach. However, the reality was that the enterprise systems were hardly up to performing all the functions in a company, especially not for the e-Factory. As a result, using a best-of-breed solution set is often the best approach to the e-factory. Thus, the "solution set" for an e-factory will likely consist of multiple suppliers.

The software and hardware companies serving the e-business and the e-factory marketplace are changing their products on a continual basis to better serve the market needs. As a result, the target set of solution candidates is not easily identified. The purpose of this chapter is to identify the leading, newest solution set providers in each of the solution set categories. While these lists will likely change as technology and application experience evolve, they do provide a starting point for an e-factory roadmap project.

Definition of a Solution Set

e-Factory solution sets are evolving from several directions. Very specialized software companies have been formed and have rapidly evolved over the last few years. They have been serving distinctive needs in separate areas in the factory as islands of automation were installed throughout the factory. Many of these small software companies have since been acquired by larger systems integration and hardware companies.

Manufacturing equipment companies that were started to serve specific process automation needs have now evolved to the point where they are trying to serve the needs of an integrated factory floor system. Process hardware companies are typically more concerned about getting the physics right in their manufacturing equipment and including only enough software to allow a human operator of the machine to be able to set up and operate the machine correctly. The technical evolution in the e-factory now requires that this equipment operate in a factory computer network and communicate with systems at higher control levels.

Systems integration companies, which initially began by providing the technical experts required to build interfaces between the different software packages and machine controllers that were necessary for a factory to operate electronically, have also begun to acquire software companies and offer solution sets for the e-factory market.

There is no off-the-shelf solution set that fits all, yet. ERP companies have advertised that they offer all the solution components that a manufacturing enterprise needs, but they do not, as yet, address the e-Factory. There are supply chain management software companies that are rapidly bringing new product features to market to address the business-to-consumer and business-to-business market needs, but they too do not yet address the total e-Factory. However, these products are rapidly evolving and it is important to be able to evaluate them when developing an e-factory roadmap.

Solution set refers to a set of software and hardware components that provides the e-factory capability. These components might come from different software and systems suppliers or be provided by one supplier. The solution set must be able to execute, facilitate, and accelerate the internal business processes and the operational scenarios that define the way an e-factory should work. These components must also support and execute the interfaces with the other internal and external processes that exist in the e-factory.

The Components of a Solution Set

The components of a solution set consist of hardware and software products whose operations cover all levels of the e-factory control hierarchy and both the vertical and horizontal dimensions in the supply chain.

As discussed in other chapters, the control levels are important because they provide a means of identifying what types of control are required, where they should be applied, and how they should be implemented. The control levels of the e-Factory hierarchy include:

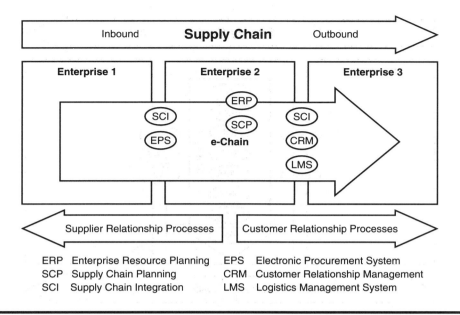

ERP Enterprise Resource Planning   EPS Electronic Procurement System
SCP Supply Chain Planning   CRM Customer Relationship Management
SCI Supply Chain Integration   LMS Logistics Management System

Exhibit 6.1   Horizontal Solution Set Components

- factory level
- line level
- cell level
- machine level
- unit level

The horizontal components of an e-factory solution set address the electronic supply chain, or "e-chain," requirements. They address the external business issues for the e-factory. Building on the enterprise and supply chain models introduced in Chapter 1, the horizontal e-chain is illustrated in Exhibit 6.1. The horizontal components include:

1. customer relationship management (CRM)
2. supply chain planning (SCP)
3. electronic procurement system (EPS)
4. supply chain integration (SCI)
5. logistics management system (LMS)

The vertical components of an e-factory solution set address the requirements that are created by the need to support the internal process requirements and the interfaces to the e-chain solutions that different customers and suppliers might have. The enterprise and supply chain model extended to illustrate vertical solution set components is illustrated in Exhibit 6.2. The vertical components include the following:

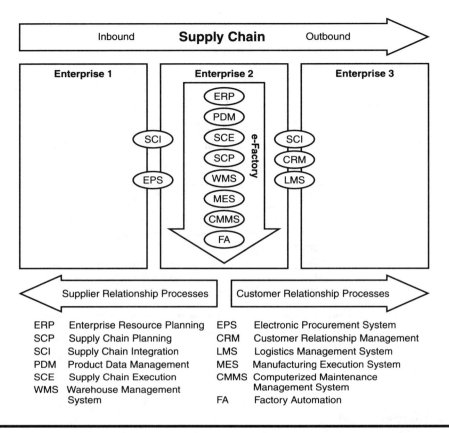

ERP   Enterprise Resource Planning   EPS   Electronic Procurement System
SCP   Supply Chain Planning   CRM   Customer Relationship Management
SCI   Supply Chain Integration   LMS   Logistics Management System
PDM   Product Data Management   MES   Manufacturing Execution System
SCE   Supply Chain Execution   CMMS   Computerized Maintenance
WMS   Warehouse Management   Management System
       System   FA   Factory Automation

Exhibit 6.2   Vertical Solution Set Components

1. enterprise resource planning system (ERP)
2. supply chain execution system (SCE)
3. manufacturing execution system (MES)
4. warehouse management system (WMS)
5. factory automation system (FAS)
6. factory floor control system (FCS)
7. computerized maintenance management system (CMMS)
8. product data management (PDM)

Horizontal Solution Set Components

Each of the horizontal solution set component categories is described in the following.

Customer Relationship Management

Customer relationship management (CRM) is a term that has evolved to describe an entire industry of software vendors that serve the specific purpose of addressing the selling and marketing functions in a business enterprise. The functions typically included in CRM applications include:

- salesforce automation
- call center management
- customer profile management
- on-line product catalogs
- data mining
- product configuration
- proposal generation
- proposal management
- order entry
- order status management

These functions are provided in separate software packages and in integrated packages. Highlighted in Exhibit 6.3 are the overlaps that exist between these different functional areas.

Salesforce automation (SFA) refers to those capabilities associated with keeping track of customers, the frequency of customer contacts, the scheduling of customer contacts, tracking and reporting on the status of sales opportunities, tracking and reporting on the sales success of each salesperson, and salesforce expense reporting.

Call center management (CCM) software performs a variety of functions necessary to automate and facilitate the activities of a call center. Companies that have a need to provide real-time information to their customers are using

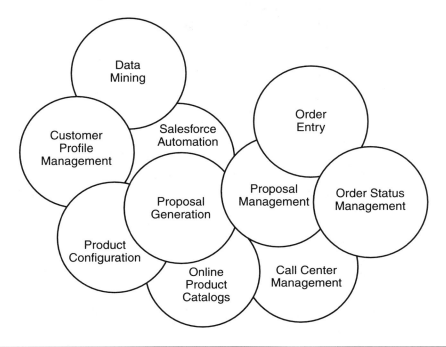

Exhibit 6.3   CRM Application Overlaps

both on-line computer technology via the Internet and "on-line" human response via the telephone. Despite the innovation and improvement of computer technology over the last few years, the flexibility and psychological reward of having a person to talk to has not be supplanted. As a result, many companies are using a combination of computer technology and people to provide a comprehensive capability to answer telephone calls from their customers.

To be effective, a call center requires a variety of software. Automated answering and routing software is now available from private branch exchange (PBX) vendors. This is the software that provides the "push 1 if you want to make new reservations, push 2 if you want to check on your air miles" type of responses with which most of us are familiar. Customer and product information database software is also necessary for a call center. The people answering the phone calls must be able to answer fundamental questions about products, pricing, and the status of a customer's account. These database systems are necessary for both the Internet on-line functionality and the call center functionality. A third type of software capability is the tracking of the performance of the people and the technology operating in the call center in terms of calls answered successfully, number of transfers required during a call, time a caller had to wait, etc. Call center management software provides these types of features as well as many of the other features described above.

Customer profile management refers to those capabilities associated with keeping track of customer purchases, complaints, preferences, personal information, response to direct selling techniques, and level of satisfaction achieved.

On-line and real-time product and service catalogs refers to making it possible for the salesperson directly or for the customer directly to be able to access a catalog that describes the products and services available from the selling company. This catalog can also include the appropriate prices and instructional information that can help a customer decide which product might be the best for a particular application.

Data mining software provides the capability to analyze a database of information or in some cases a continuous flow of information to identify key trends, relationships, and behavior of individuals and groups. This software a few years ago was touted as a fantastic breakthrough in marketing because it had the promise of providing "scientific" insight into how people's shopping behavior indicated what they might buy in the future. The concept is to apply advanced statistical analysis to very large databases of shopping activities and customer demographics to imply relationships between products and current customers. Thus, a person who goes to a grocery store to buy disposable diapers for a child and who pays for the item using a grocery store discount card becomes a database record. The record would include their name, address, telephone number, an estimate of their annual income, the number of times they buy diapers, other baby products they might buy, etc. At some point in time, this information would be processed by a data mining software package that would put the person's name and address into different mailing lists that advertisers of baby-related products, vacation package retailers, and others would use to begin sending catalogs and direct selling activities.

Proposal generation refers to those capabilities associated with generating a proposal to reflect the needs of the customer. This capability should include the ability to configure a proposal from a set of predetermined subsets of activities in the same way that one might create a meal and therefore the price of a meal from a menu.

Proposal management refers to the status tracking of a proposal once it is submitted to a customer for decision. Proposals are in many ways like fresh fruit in that they tend to diminish in quality with time until after a certain point they expire and are no longer viable.

Order entry is the creation of an order once a proposal has been accepted by a customer. This has traditionally not been part of the customer relationship management set of functions but it is becoming more so now that on-line purchasing is becoming more acceptable.

Order status management is the management and reporting of the status of a customer order. For build-to-order businesses, this function is quite important because it allows customers to track and understand the progress of their order. For ship-from-stock businesses, order status management is necessary as a real-time performance indicator of the quality of their order fulfillment service.

Although a list of companies that offer CRM capabilities is given in the appendix, the leading multi-purpose CRM providers include the following companies.

- Siebel Systems, Inc.
- Oracle Corp.
- PeopleSoft, Inc. (Vantive)
- Trilogy Software, Inc.
- Nortel Networks Corp. (Clarify)
- Baan Co.
- Davox Corp.
- Aspect Communications, Inc.
- Astea International, Inc.

Supply Chain Planning

Supply chain planning (SCP) for a manufacturing company generally refers to long-range planning of the design and operation of the supply chain to which the company belongs. Supply chain design refers to the number of plants, the capacity of the plants, transportation modes, outsourcing options, the number and type of suppliers, the types of contracts with the suppliers, the number of warehouses and distribution centers, and the flexibility in assignment of each component in the supply chain. Supply chain operation refers to the number of shifts in each plant, decision rules for shipments and transportation, replenishment rates, and cycle times of each component in the supply chain.

Very long-range or strategic supply chain planning is done only infrequently, either on an ad hoc basis or on a regular or annualized planning basis. The ad hoc basis can be driven by events such as new product developments, consolidation of merged business units, or financial crisis. The regular or annualized basis is often part of a regular strategic planning schedule. As part of the longer-term perspective, supply chain planning tends to be focused on the structure of the supply chain and the simulation of different operational strategies. For the e-factory, this perspective becomes very critical because all new supply chain designs must be able to perform within the short cycle times and with the flexibility being required by the marketplace.

Short-term supply chain planning is done on a more frequent basis, which could be quarterly, monthly or weekly. The e-factory environment is forcing supply chain planning to be done even more frequently. At some point, what is called supply chain planning becomes identical to supply chain execution. Short-term supply chain planning focuses on activities that are more operational or low-level, such as multi-plant production scheduling.

A key part of supply chain planning is the search for the optimum plan. Optimization includes deciding on the location, size, and the number of plants, distribution centers, and suppliers that provide the best performance for the supply chain. Of course, performance must be defined as part of the supply chain planning process. Supply chain planning also includes sourcing and deployment strategies for each plant, each distribution center, and each customer. It also models the flow of goods through the supply chain network.

Developing a supply chain plan for one product is an interesting mathematical and management challenge. Expanding the plan to include multiple products makes the mathematical challenge exponentially greater. Expanding the plan to include multiple plants takes the mathematical challenge to an even greater level of complexity.

The leading supply chain planning companies include the following.

- i2 Technologies, Inc.
- SAP AG
- Manugistics Group, Inc.
- Aspen Technology, Inc.
- Baan Co.
- LPA Software, Inc.
- Logility, Inc.
- SynQuest, Inc.
- PeopleSoft, Inc.
- ILOG, Inc.
- Intentia International

A longer list of companies is included in the appendix.

Electronic Procurement System

When the operations of a manufacturing company are analyzed from a supply chain perspective, the basic question that results is how much value is being added by the manufacturing company to the goods and services that it purchases to produce a product that is sold to its customers. For many industries, the amount of purchased goods and services used directly in the manufacture of a final product runs in the range of 60 percent to 90 percent of the final manufactured cost of the product. In gross terms, the cost of purchased goods and services can run in the range of 25 percent to 50 percent of total revenue. From this follows the obvious observation that any decrease in the cost of purchased materials is amplified in the increase in net income because it falls to the bottom line directly. For example, if the cost of purchased goods and services is 80 percent of the cost of the manufactured goods and if the company sells the product so that it has a 50 percent gross margin, then the cost of purchased goods and services is 40 percent of revenue. If the cost of purchased goods and services is reduced by 5 percent, then that savings (which is 2 percent of revenue) drops directly to the bottom. If this example company was achieving 10 percent of revenue after tax earnings before, then the earnings would increase to 12 percent. Thus, the 5 percent decrease in purchase costs translates into a 20 percent increase in earnings. This is the dream that all CEOs have when they hear about the promise of electronic procurement systems. And some vendors and consultants promise 20 percent reductions in purchased material costs, which translates into net income percentage increases that are too exciting even for dreams. For some companies in slow growth industries, cost reduction of purchased goods is the only strategic option they have to show profit increases. The concept of using software to improve the earnings of a company, which has been promised before by virtually every new wave of business software developed, is almost too good to be true.

There is some substance, however, to the potential value of electronic procurement systems (EPS) to assist in the reduction of purchasing costs. The benefit comes primarily from two sources. First, the procurement business process in most companies today is inefficient in terms of the execution of purchase decisions and transactions, and wasteful in terms of matching purchasing needs with purchasing reality. EPS can reduce the transaction cost of the procurement business process. Second, there is now a body of knowledge and best practices on how to minimize the cost of purchased materials by combining purchasing requirements to buy at larger volume levels and by designing cost out of products by working more closely with suppliers during product development projects. EPS can help make better decisions about cost by searching through electronic catalogs for the lowest cost components that offer the same functionality (i.e., comparison shopping).

There are several ways for companies to execute purchasing decisions, including:

- buy at an auction
- buy from a distributor
- buy from retail stores or shopping centers
- buy from a supply chain partner

EPS can support all of these forums. They can also assist in deciding from which forum to purchase goods and services.

The basic elements that form an EPS are simple.

- All potential suppliers product catalogs are electronic and on-line.
- All terms and conditions (shipping quantities, prices, freight costs, warranty, etc.) in all purchase agreements between company and suppliers are on-line.
- All purchase requests, ordering, status checking, and receiving activities are an on-line and available to everyone that makes the final decisions.
- All settlement of purchase transactions are electronic.

The result is that all the inefficiencies in time and labor that come from having people in functionally specialized departments performing the purchase request form processing, shopping, buying, and settlement activities can be eliminated with automation. However, the cost of the procurement process (i.e., the cost of the people, facilities, and equipment employed within the procurement process) is small in comparison to the cost of the purchased materials. If purchased material is 40 percent of revenue, the cost of the purchasing department is 0.4 percent of revenue. The big cost saving is not going to come from the replacement of manual processes by automated processes. If fact, the cost of the automated process in terms of software and hardware costs could exceed the cost of the manual processes being replaced.

In addition, the new contractual relationships between supplier and customer that must be negotiated in light of the new e-business world will require a new, enlightened perspective on business processes, both internal and external, and on how supply chain partners will operate. This will require more talent and energy than the traditional types of buy/sell contract terms and conditions. Thus, the cost of establishing a contractual relationship may increase, especially if the new relationship requires suppliers to have a certain level of e-business capability to support the supply chain relationship being negotiated.

Then there is the issue of getting all product catalogs on-line. The suppliers can invest in putting their catalogs in electronic form rather than in paper form. In fact, with the tools that most companies are using today for electronic publishing, getting product catalogs on-line can be done as part of a multimedia product catalog strategy. The question then becomes one of standards and interfaces. The more standards and interfaces that a company has to support, the more cost incurred.

The real promise and benefit in electronic procurement is in reducing the cost of purchased materials through better decision-making. EPS can make a major contribution to this with the proper understanding of how a procurement decision process works and how it should be automated or supported.

There are two general types of procurement activities in most manufacturing companies: those for direct manufacturing and that for maintenance, repair, and operations (MRO). Effort has been focused on purchasing of goods for direct manufacturing from many quarters. MRP, PDM, ERP, and SCM packages are all designed to make the material planning and procurement processes more responsive to corporate needs. The introduction of the external business process concepts and supporting software have helped as well.

MRO buying is a relatively new area of opportunity and is the area that has received the most attention from electronic procurement software companies that have started up over the last five years and the companies that are buying their products. MRO includes everything from replacement tooling in a manufacturing facility to copier paper in the human resources office. In fact, MRO buying is a significantly more important function in service industries because there is no manufacturing material being purchased. MRO buying in the financial, insurance, education, government, and entertainment industries consumes larger percentages of the revenue than in manufacturing companies, and there has historically been less money invested in improving the procurement process in those industries. Thus, there are many new opportunities in MRO.

Specialized Web sites called e-markets and vertical portals have been created where people with similar business interests and needs can trade and talk business-to-business (B-to-B) on a round-the-clock basis. Suppliers, ranging from office products vendors to very specialized producers of raw materials or manufacturing components, have opened Internet stores and B-to-B marketplaces that post their catalogs and offer other client services.

A 70-year-old distribution company, W.W. Grainger, took its various 220,000 products to the Internet. At Grainger.com, a company can establish a corporate account, authorize purchasing agents with specific capabilities, review the status of accounts and shipments at any time, and receive one consolidated bill. Grainger's newest site, OrderZone.com, streamlines customers' purchasing processes across multiple suppliers. Office product vendor Staples has started up StaplesLink.com, a site that lets corporations open accounts and receive discounts on orders. US Office Products, will sell you everything from paper clips to coffee machine services, as well as consolidate your accounts.

Another innovation is online services that route purchase requests to appropriate managers for approval based on policies, provide customized online catalogs, and generate reports to help you track spending and negotiate volume discounts. Electronic and online procurement systems can have customized links to accounting, project management, and other resources. The enterprise system never has to touch the Internet, so internal security can be very tight.

The area of e-procurement is evolving rapidly. Ariba and Grainger are collaborating on specialized features written in Commerce XML (CXML) to automate the handling of supplier content. The suppliers receive orders by e-mail, fax, or electronic document exchange and settle accounts by electronic links through banks. In the most basic form, buyers need only an Internet connection, a browser, and a password to browse the combined material.

In its ultimate form, an e-market can become part of an enterprise's supply chain management system. Companies using ERP systems, such as those from Baan, J.D. Edwards, Oracle, PeopleSoft, or SAP, to control production, inventory, and accounting can benefit from using special e-market ordering programs tied to their ERP systems. Supply chain management, integrated with ERP, provides an automated way to order materials and settle accounts. It also supplies performance data on order fulfillment times, rejects, and other statistics that can give procurement professionals an objective way to measure vendors and the confidence to lower inventories and optimize production schedules.

The leading EPS vendors include:

- Ariba
- CommerceOne
- Trilogy Software
- i2 Technologies

A more complete list is included in the appendix.

Supply Chain Integration

One of the rapidly evolving horizontal applications in the e-factory is the area of integration between the internal and external business processes. This area is known under several names (e.g., enterprise application integration (EAI) or cross-applications), but the term "supply chain integration" (SCI) will be used herein. This area includes generic software services for security management, protocol management, data mapping, and software connector modules that let internal process software communicate with external process software.

Most of the SCI software packages are designed to use technologies and standards that incorporate one of the following: XML as a data-interchange format, Enterprise JavaBeans as a server architecture, CORBA or DCOM as programmatic methods for processes to call each other, message queuing systems to let systems communicate asynchronously, and transaction monitors to ensure that operations complete properly.

There are various types of supply chain integration. Typically they are application-to-application integration and customer-to-business integration. SCI software provides data transformation, mapping, workflow technologies, graphical tools to display data as seen by various applications such as ERP systems, transaction monitors, and databases. XML is fast becoming the unifier among integrated systems. XML is also the basis for SOAP (Simple Object Access Protocol), a new way for processes on different systems to communicate using standard, open protocols. SOAP could greatly simplify interoperability among systems because of the openness and simplicity of XML. SOAP is a communications protocol, not a description of the data in an e-business system. The fact that SOAP can run over any transport protocol (e.g., HTTP, SMPT, or message queuing) means that any two SOAP programs can communicate without another layer of middleware.

CORBA/IIOP and DCOM are the two main protocols for object communication across networks, including the Internet. These are low-level programming concepts, but often dictate compatibility between applications in a complex system. DCOM is used for Windows-based applications, and CORBA is found on many platforms. Because both have wide support in the market, one finds that many EAI systems, including those from Bridges for Islands and Saga Software, support both.

Not all calls between programs need an immediate response. Message queuing is a long-established technique that lets programs call each other using connections over which responses are not necessarily immediate — Internet connections, for example. Message queues deal with these calls asynchronously and can resend them if necessary. Message queues are an excellent means of connecting disparate systems, especially across the Internet.

The leading software vendors include the following.

- Applix
- Coral Sea Software
- Maconomy International Software
- IMAJ
- BroadVision
- QAS
- Saratoga Systems
- Silknet
- CrossWorlds
- FileNET
- Extricity

Logistics Management

Logistics management is the management of the warehousing, storage, transportation, and material tracking activities. The general area of logistics management has actually evolved independently from each of these specific areas and is merging into a unified solution set component. The trends in the e-factory are causing fundamental changes in the way logistics is managed and in the way companies think about logistics in their strategic structure.

At the highest level, every e-factory must be concerned with logistics management because of the need to ensure that the physical delivery portion of the order fulfillment process is successfully completed. As a minimum, information about material tracking, scheduling, and delivery confirmation is important to the e-factory. This information might come from an external process because the physical logistics operations have been outsourced or from an internal process because physical logistics operations are part of the e-factory core competency.

At the lowest level, the logistics process must be constructed to ensure that goods are stored, warehoused, picked, packed, shipped, and delivered as efficiently and as rapidly as possible. Volume and speed are critical parameters in determining success in a logistics process. As a result, full-service logistics companies have emerged so that they can offer the efficiencies of scale in logistics management to both large and small e-factory enterprises.

The leading logistics management software companies include:

- Manugistics, Inc.
- TRW, Inc.
- J.D. Edwards & Co.
- McHugh Software International
- Manhattan Associates, Inc.
- EXE Technologies, Inc.
- Optum, Inc.
- Catalyst International, Inc.

Vertical Components

Enterprise Resources Planning Systems

During the past decade, no single software industry segment received more visibility, experienced more growth, was sought after more by customers, was stimulated by more experts and non-experts, and caused more confusion than that of enterprise resource planning (ERP) systems. What started out in the decade as the arrival of a promising new business system concept that employed the latest in mainframe computer technology ended up as a millennium bug solution fad that was being touted by the news media, high government officials, and the mainstream press.

In reality, ERP systems were a logical evolution in the business systems industry during the last ten years because there was a need for a business to integrate its operations using networked computers rather than a single mainframe with a myriad of terminals hardwired to it. However, the theory that a single software system could provide every functionality needed by a manufacturing company and also provide major productivity improvements just from the sheer force of technology was, in fact, difficult to accomplish. The implementation effort required to "configure" a generic ERP system and implement it in companies with multiple plants, legacy systems, established databases, and business processes in need of improvement often approaches or exceeds the effort required to build a custom system from a library of application modules or objects that are custom developed or designed for standard application.

The other problem with ERP implementations has been that of every major new system conversion: change management. The changes caused by the introduction of new system technology posed major new problems that had never been seen before. There were many reasons; but one of the biggest reasons was that the new ERP technology was designed to be used by virtually everyone in the company to record and report all the transactions and activities in the company. Thus, the change of systems technology was compounded by the fact that many employees who had not had to deal with computers as part of their workday life suddenly had to depend on them for every major action they took. It has taken some companies years to overcome the disruptions in operations due to the introduction of ERP.

All that having been said, the reality is that the concept and practice of enterprise solutions is here to stay. As stated elsewhere, one of the key issues is whether the enterprise "solutions" should be monolithic from one supplier or best-of-breed from several suppliers. The objective of this book is to provide a methodology and tools for answering this question as intelligently as possible for the situation at hand.

Despite the actual track record — which is not one of total success — enterprise software products have evolved and improved and do represent an alternative or choice for many aspects of the e-factory requirements. While the term "enterprise" is a misnomer because these software systems do not provide all requirements for an enterprise, it does represent a significant genre of software packages. The leading enterprise provide are as follows.

- SAP
- Oracle
- Baan
- J.D. Edwards
- PeopleSoft

These leading ERP suppliers are continuing to invest heavily in expanding and improving their products to make them easier to integrate into a diverse solution set environment.

Supply Chain Execution Systems

Supply chain execution systems for manufacturing generally refer to systems that execute the transactions necessary for the supply chain to operate successfully. Supply chain execution systems include warehouse management, transportation management, logistics management, and distribution management systems.

The e-factory-based supply chain must respond to the needs of the Internet economy. The e-factory must be quicker, smarter, more flexible, and be able to process everything from an individual production unit to containers and truckloads. Supply chain execution systems are therefore rapidly evolving to support these needs. As a result, there is a blurring of the supply chain planning and supply chain execution systems packages.

Prior to e-business, manufacturers would have to plan to ship truckloads of pre-built product through retailers to corporate accounts. Now, manufacturers use distribution centers for custom configuring, assembling, and shipping smaller, even individual, orders. As a result, manufacturers must sell more individual orders to make the same revenue once generated from fewer but larger orders.

New companies are being formed which provide software and Internet-based services that are completely changing how manufacturers view their business and manage transportation. Manufacturing companies can now get everything they need from a software package, an Internet exchange, or an outsourcing company. Benefits include the ability to measure a carrier's performance and gain control over maverick transportation spending by negotiating volume service agreements.

Successful LX portals will offer transportation management software, provide visibility to shipments, and supply connectivity and supply chain event management to coordinate the activities of an entire trading community.

Leading supply chain execution system vendors include:

- Interactive Business Systems AB
- J.D. Edwards & Co.
- McHugh Software International
- Manhattan Associates, Inc.
- EXE Technologies, Inc.
- Optum, Inc.
- TRW, Inc.
- Catalyst International, Inc.
- Provia Software, Inc.
- Descartes Systems, Inc.
- Swisslog Management AG
- STS, Inc.
- HK Systems, Inc.
- Vastera
- Industri-Matematik International Corp.

Manufacturing Execution Systems

Toward the end of the 1980s, several experiments were conducted in the automotive, aerospace, and consumer goods industries in automating a manufacturing environment using custom-developed computer software. The results were promising enough to create a market for a general-purpose system that would monitor and control the activities and equipment in a manufacturing process. As a result, the term "manufacturing execution system" (MES) was coined in the early 1990s. It was created to define the software evolving at that time, which included a variety of functionality that associated with the control, monitoring, and reporting of factory floor processes. As with everything software, the functionality over the last ten years has expanded dramatically and now overlaps with other categories of e-factory software. A list of functionalities that are considered part of the MES category follows.

1. Operations monitoring
   a. Monitoring and tracking
   b. Product and workflow management

      c. Performance analysis
      d. Production reporting
      e. Cost accounting
      f. Labor tracking
      g. Finite scheduling
      h. Schedule dispatch
      i. Maintenance management

2. Manufacturing process engineering tools
      a. Process simulation
      b. Process engineering development tools
      c. Process planning (CAPP)
      d. Electronic document management

3. Quality management
      a. Quality documentation
      b. Audit and defect tracking
      c. Supplier management
      d. Statistical process control
      e. Inspection and test
      f. Laboratory information management systems
      g. Statistical analysis
      h. Cost of quality

4. Process control
      a. Supervisory control
      b. Cell control
      c. Man–machine interface
      d. Data acquisition control
      e. Historian and archive
      f. Recipe management
      g. Numerical control
      h. Operator support
      i. Data collection

The leading MES software suppliers include:

- Emerson Electric (Intellution)
- Invensys (WonderWare)
- Brooks Automation (FASTech)
- Applied Materials (Consilium)
- Hilco Technologies
- Real World Technologies
- POMS, Inc.
- USDATA

Warehouse Management Systems

A key part of the e-factory is the rapid and efficient control and movement of the material that is purchased and created by the manufacturing process.

Finished goods warehouses have employed material handling automation for many years. As supply chain cycle times shrink, the size of shipped lots shrink, and the mix of items in an order becomes more diverse, the complexity of managing a warehouse increases to the point where the control can only be done by a powerful computer software system. Warehouse management systems were brought to market during the past ten years. The features offered in warehouse management systems include the following.

1. Quality control
2. Reporting
3. Returns
4. Transportation management
5. Order tracking
6. Fulfillment notifications
7. Receiving
8. Storage
9. Order selection
10. Loading
11. Shipping
12. Inventory arrangement

The leading warehouse management systems suppliers include the following.

- EXE Technologies, Inc.
- Catalyst International, Inc.
- JDA Software Group, Inc.
- Manhattan Associates, Inc.
- McHugh Software International
- Optum, Inc.
- TRW, Inc.

Factory Automation Systems

Factory automation systems (FAS) refer to the various hardware and software components, equipment, and systems that form the actual link between the material, processes, and information in a production line. Such items include:

1. Programmable logic control (PLC)
2. Robotics
3. Material handling systems
4. Cell controllers
5. Computerized numerical control
6. Intelligent valves
7. Actuators

The leading suppliers of factory automation systems include:

- ABB Industrial Automation
- Siemens Energy & Automation, Inc.
- Invensys plc
- Fisher-Rosemount Systems
- Honeywell
- Rockwell Automation
- Yokogawa Electric Corp.
- Cutler-Hammer
- GE Fanuc Automation Corp.
- Schneider Automation, Inc.

Factory Floor Control

Factory floor control systems (FCS) originated with the need to control high-volume, process-intensive production runs in industries such as chemicals, food, pharmaceutics, and automotive. The advent of e-business brings with it the requirements of having lower volume, more frequently changed, higher quality, build-to-order production processes. This puts more pressure on factory FCS to be able to facilitate and execute the actions required to have such capability.

Trends in process control include: (1) componentization of software that enables capabilities to be tailored and flexible; (2) predictive tools that permit real-time process optimization; (3) shrinking production cycle times that spur interest in process automation; and (4) e-business flexibility that demands plant floor visibility.

With optimization and connectivity to suppliers and customers as goals, factory floor control software is bringing to the table more sophisticated modeling capabilities and predictive tools, as well as enhanced data analysis and more robust management of infrastructure and production. As a result, companies are seeing improvements in information flow, asset management, operational flexibility, process efficiency and reliability, and throughput. Improvements are both initially high, in the capital investment phase and in the long term, as productivity and quality improvements reduce expenses.

While important, savings alone will not persuade a company to invest in process execution and control software. The software must also be easy to use. Ease of use is important for two reasons. First, there is a shortage of IT professionals, and qualified consultants are expensive. Second, IT professionals have been trained to use off-the-shelf office applications and do not have the time or patience for complex installation sequences and long learning curves. Products will have to be much more plug-and-play than process industries have seen in the past.

For users in traditional and new markets, the availability of modeling tools is increasing the appeal of factory FCS. As optimization has grown in importance, plant modeling has become a desirable feature. Six Sigma quality programs have clearly proven the need for quantitative analysis. Six Sigma quality means 3.4 defects per million opportunities. To achieve these high quality goals, significantly more accurate and reliable data is required for analysis.

The top ten suppliers for factory automation include the following.

- Honeywell
- Fisher-Rosemount Systems
- Invensys plc
- Rockwell Automation
- GE Fanuc Automation Corp.
- ABB Industrial Automation
- Aspen Technology Inc.
- GSE Systems, Inc.
- Nematron Corp.
- USData Corp.

Maintenance Management Systems

Computerized maintenance management systems (CMMS) or the newer term of "Enterprise asset management" (EAM) systems refers to the software that provides the management and control over the utilization, repair, and tracking of the equipment used in a manufacturing facility. Maintenance management software can extend equipment life; reduce downtime, repair costs, and capital investment; increase productivity; and improve product consistency.

The history of CMMS goes back many years to the earliest days of computer usage in manufacturing. The functions required in managing the maintenance for a large factory have always included:

1. Inventory management of the equipment to be maintained
2. Tracking of the scheduled and unscheduled maintenance performed
3. Maintaining and creating a database of suppliers of repair parts and services
4. Purchasing repair parts and services
5. Monitoring equipment availability and performance status

Stand-alone computers have always been used to store and manage the above information. New requirements for which there are available capabilities include:

1. Condition-based maintenance
2. Real-time equipment performance monitoring
3. Remotely managed maintenance schedules
4. Remote diagnostics analysis
5. Real-time online purchasing of repair parts and services

In addition to the software available, virtually all EAM software vendors have begun to offer or are working on online hosted models, which many view as more affordable and easier to implement. Other in-demand features include ease of use, real-time communication, and portability. Using these types of systems workers rely on wireless handheld devices to access work order assignments e-mailed from the program. Upon completion of a task, the worker reports back, detailing time spent and corrective actions taken. This input can be tracked to provide a snapshot of work in progress.

The leading maintenance management systems include:

- Indus International, Inc.
- PSDI
- Datastream Systems, Inc.
- Invensys plc
- Meridium, Inc.
- Mincom, Ltd.
- Fluor Daniel, Inc.
- Ivara Corp

Product Data Management Systems

Product data management (PDM) systems have evolved from the need to create, store, distribute, update, reuse, and glean information from data associated with the design and manufacture of products. These tools accomplish this by managing workflow as well as data. The primary objective of PDM systems is to increase the productivity and efficiency of the product realization process. However, additional objectives are being added as the needs of the e-factory grow.

The objectives and expected benefits of PDM include:

1. Classification control
2. Product structure
3. Workflow management
4. Work history management
5. Improved design productivity
6. Safeguarded data integrity
7. Intellectual property management
8. Product structure management
9. Change management
10. Configuration control
11. Automatic data release
12. Electronic sign-off procedures
13. Revisions
14. Audit trail of changes
15. Collaborative product realization

Collaborative product realization is one of the fastest growing new requirements and promises for PDM. Because of enterprise fission and supply chain fusion, the need for and the opportunity for intensive collaboration across enterprise boundaries are increasing. The key features of PDM that have been under development and enhancement for the past 20 years are now available online and in real-time. As a result, collaboration in the e-factory environment is one of the essential ingredients going forward.

Collaboration through PDM, also called collaborative product commerce (CPC), pays multiple rewards. Engineers are often conservative in their approach to problem-solving for no other reason than the time penalties for exploring alternative solutions are so high. PDM opens up the creative process by keeping track of all the documents and test results. This minimizes design rework and mistakes, reduces the risk of failure by sharing the risk with others, makes data available to the right people fast, and encourages team problem-solving and collaborative idea generation.

Leading PDM vendors include:

- SDRC
- PTC
- Documentum, Inc.
- MatrixOne, Inc.
- Unigraphics Solutions, Inc.
- Enovia Corp. (Dassault Systems)
- Eigner + Partner AG
- Agile Software Corp.
- SmarTeam, Inc.
- Auto-Trol Technology Corp.

Summary

The e-factory environment is now served by hundreds of software and hardware products that represent a rich and diverse portfolio of solution components from which a solution set can be selected. The task of selecting a solution set depends on first understanding the business process needs, matching the capabilities of the candidate solution components to those needs, and then selecting the best set of components based on a well-defined set of performance criteria. This chapter has introduced the variety of solution components available on the market and where they serve key business process needs.

Chapter 7

Visioning the e-Factory

Purpose of This Chapter

A roadmap describes a path from a starting point to a destination. The destination for the e-factory roadmap is an e-factory operating environment that consists of an integration of process and technology. One of the objectives of an e-factory roadmap is to define a clear path to a process and technology destination. Because the e-factory is a new and rapidly evolving facet of e-commerce, creating the roadmap will likely include discovery of internal needs and external requirements as well as definition of which turns to take when the proverbial "fork in the road" is encountered. One of the fundamental questions that will often require discovery is the determination of the eventual destination. While the roadmap methodology described in Chapter 8 is designed to assist in the discovery process, the overall effort to get to the detailed definition of an e-factory can be optimized if there is an initial vision of the destination. The purpose of this chapter is to describe how to obtain an initial vision and what should be included in such a vision.

A vision should include the definition of the goals, direction, and urgency that accompany the creation of an e-factory. A vision provides a soft, fuzzy image of what the cold, hard details of a solution set might be. The visioning process is also a way to facilitate gaining the consensus of a group of stakeholders in the e-factory on what should be accomplished.

The following paragraphs define the contents of an e-factory vision, examine the differences between top-down and bottom-up visioning, and identify the three sources of requirements for the vision.

The Key Elements of an e-Factory Vision

The vision of an e-factory needs to be robust enough to include all of the key elements that provide the flexibility and speed embodied by an e-factory.

These key elements contribute to the total fabric of the ultimate vision while remaining important in their own right. There are five key elements of an e-factory vision:

1. Process flow vision
2. Process boundary vision
3. Control vision
4. Technology vision
5. Change management vision

These key vision elements are important not only to the total e-factory vision, but also to the analysis required to define the e-factory roadmap.

Process Flow Vision

In general, the manufacturing environment is a dynamic, multi-dimensional, multi-media system with currents of activity flowing in different layers with both smooth and turbulent flow characteristics. These currents of activity typically include the flow of material, the flow of work, and the flow of information. These flows are moving from sources to receivers and are sometimes intertwined and sometimes totally disconnected. In fact, the amount of turbulence in these flows is a good indicator of the efficiency and effectiveness of a particular factory.

A illustrative diagram of these different flows is given in Exhibit 7.1. The process flows are either a byproduct of good advance planning that occurs in the design of a business process, or the result of random chance that occurs when a process is just thrown together and left to evolve. One traditional

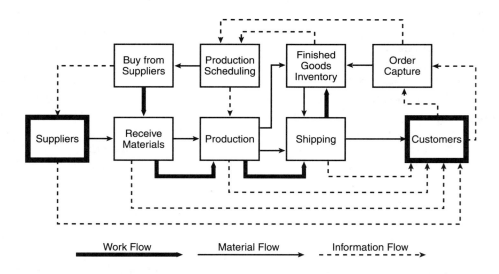

Exhibit 7.1   Multiple Flows in the e-Factory

approach to factory layout design has been to (1) identify the manufacturing equipment needed to produce the product, (2) decide on a layout of the equipment, (3) identify the staffing requirements, (4) buy all the equipment, (5) install the equipment, (6) begin production, and (7) identify and fix start-up problems. Too often, however, one or more of the three types of flows (i.e., material, work or information) are either overlooked until production is attempted or are not thoroughly defined and tested. Understanding what each type of flow requires and how each should operate is important for factories... but is essential for the e-factory.

Workflow is the movement of tasks, responsibilities, and action items through an organization structure and facility. The flow of work represents the philosophy of the management approach as well as the efficiency of the business process design. Software that facilitates and implements workflow management throughout an organization has been commercially available as an independent package for several years. Most enterprise systems and many e-business applications have some form of workflow management capability. The most popular e-mail systems today can operate in a workflow management mode.

Workflow management is important in the e-factory because of the need to have fast but flexible business processes. The need for speed comes, of course, from the fact that the electronic media is being used to communicate the flow control and the fact that powerful networked computers are being used to manage the business process and the relevant databases. The need for flexibility comes from the need for business processes to be modified to reflect new customer requirements.

The vision for workflow thus reflects the entire thrust of a management philosophy about how work is supposed to be done, the degree of urgency imposed on certain activities, how much discipline is required in a business process, and the level of control to be applied to the business process. The vision for workflow should include — at least — the following:

- Which resources (human, machine, or data) are to be involved in the workflow
- What performance parameters should be used to measure workflow effectiveness
- Which technologies are required to facilitate the workflow

Material flow is the movement of material (either raw material, work-in-process, or finished goods) through a facility. The course that material takes through a facility represents the philosophy of the process approach as well as the efficiency of the business process design. Over the past ten years, debates between competing philosophies on the best flow of material through a production facility have raged from continent to continent, from culture to culture, and from consulting fad to consulting fad. However, there has been a convergence of these philosophies to a consensus on the best approaches for designing and managing the flow of material. These converging concepts come from process improvement practices that use terms such as just-in-time, continuous flow production, lean manufacturing, and Six Sigma.

Software that plans and monitors the flow of material throughout an organization has been commercially available as independent and integrated packages for several years. Ranging from the MRP III systems from the 1980s to the factory floor control systems of today, technology has been used to track, monitor, and route material through a facility for the sake of optimization.

Material flow management is important in the e-factory because of the need to have fast but flexible order fulfillment processes. The need for speed comes, of course, from laying out the factory so that material spends as little time as possible being idle and the maximum amount of time having value added to it. The need for flexibility comes having the equipment associated with the material flow organized to have the shortest time.

The vision for material flow then reflects the entire thrust of a manufacturing philosophy about how production is supposed to be done, the cycle times of the order fulfillment processes, how inventory replenishment policies, and the levels of control to be applied. The vision for material flow should include at least the following:

- Which resources (human, machine, or data) are to be involved in the material flow
- What performance parameters should be used to measure material flow effectiveness
- Which technologies are required to facilitate the material flow

Information flow is the movement of information through machines, business processes, and organizations. The course that information takes through an organization represents the strategy of the process approach as well as the efficiency of the business process design. As recently as just the last decade, the bulk of information flow through a factory took the form of paper. These paper travelers, hot sheets, or kanban tags contained all the information that was necessary to identify what work needed to be done and where the material was going next. However, the e-factory depends on technology to support, facilitate, and execute the flow of information.

The types of information that flow through an e-factory obviously can differ, depending on the industry and the process design of the company. However, the e-factory should be able to provide information about customer orders, machine-specific product recipes, quality statistics from the actual processes, accumulated cost, elapsed time, and location in the facility. The information not only has to be processed, managed, and stored, but data has to be collected from the physical world and reported back to human or machine resources.

Information flow management is important in the e-factory because of the need to have fast but flexible order fulfillment processes. The need for speed comes, of course, from having the information immediately available and immediately collected. The need for flexibility comes having the storage capacity, fast access, data structures, and decision trees in place to ensure that as much of the information that would be needed by the order fulfillment process is immediately available.

The vision for information flow then reflects the entire thrust of a management philosophy about how information is supposed to facilitate business process efficiency and how business processes are supposed to generate and use data. The vision for material flow should include at least the following:

- Which resources (human, machine, or data) are to be involved in the information flow
- What performance parameters should be used to measure information flow effectiveness
- Which technologies are required to facilitate the information flow

Because there are three distinct flow dimensions operating in an e-factory environment, the logical question is whether and how these flow dimensions affect or depend on each other. As will be discussed in Chapter 8, there is a very convenient and simple technique, called object-oriented scenarios, for determining the details of the interaction and interdependence of these three phenomena. An object-oriented scenario is simply a series of statements that describe what happens as work, material, and information flow through a business process. A key concept in this scenario approach is that a business process is a set of transactions or activities that are performed by and between entities in the process. Each entity is an object that could be a person, a database, a software package, or a machine. The scenario defines what happens between the objects as the process operates.

Workflows have typically been described by organization charts, Venn diagrams, or high-level programming languages. Material flows have typically been described by production block diagram flow charts or computer simulations. Information flows have typically been described by data flow diagrams, functional block diagrams, or specification documentation. A useful way to tie all of these dimensions together while maintaining the integrity of the individual diagrams is to build the scenarios using a form called a scenario matrix. Exhibit 7.2 illustrates how a scenario matrix can bring together the key aspects of the three flow dimensions in an e-factory.

The matrix is a two-dimensional array of information that documents a variety of items. The vertical dimension documents the list of activities that compose the scenario. Scenarios should be developed by a team of process experts to ensure that all the key activities are identified. As each activity is defined, the type of interaction between the objects (Obj 1, Obj 2, ..., Obj N) is described in words and graphically by an arrow. The arrow begins with the object that originates work, material movement, or information; and ends with the object that receives the work, material, or information. The remainder of the matrix includes space for capturing information relevant to technology, cost drivers, and resource attributes.

Scenario	Resource						Flow Type	Action	Material	Cost	Cost Driver	Driver Value
	Obj 1	Obj 2	Obj 3	Obj 4	***	Obj N						
Activity 1	⟶											
Activity 2		⟶										
Activity 3		⟶										
Activity 4			⟶									
Activity 5		⟵										
Activity 6			⟵									
Activity 7	⇢											
***												
Activity M		⇢										

Work Flow ⟶   Material Flow ⟶   Information Flow ⇢

Exhibit 7.2   Example of a Generic Scenario Matrix

Ultimately, a process flow vision should include the following.

- Definition of the high level business processes
- Which resources (human, machine, or data) are to be involved in the process flow
- What performance parameters should be used to measure process flow effectiveness
- Which technologies are required to facilitate the process flow

Process Boundary Vision

Once the process flow vision is established, the next step is to develop the process boundary vision. The process flow vision described above will result in the definition of the high-level business processes that will be part of the e-factory. These processes are typically created in order to satisfy the requirements that come from three sources: customers, suppliers, and internal operations. The business processes that are part of the process vision must be defined in the context of these three sources of process requirements. Using the enterprise and supply chain models from the Chapter 1, Exhibit 7.3 illustrates how business processes are part of the individual sources and interactions between the sources.

Using this supply chain enterprise model, it is clear that there are interactions between business processes in the different enterprises. In the non-electronic commerce world, these interactions take place through phone calls, postal mail, e-mail, overnight packages, contract documentation, and personal visits. In the e-factory world, such interactions take place via direct computer-to-computer communications, user-to-Web site communications, electronic procurement system transactions, electronic exchange transactions, and/or electronic orders from multi-plant scheduling supply chain management software. Each business process may have its own controlling software system. These types of interactions raise the stakes and the importance of having a clearly defined set of process interfaces

In the world of the e-factory, being able to respond to customer-imposed processes and process interfaces is becoming a key requirement. As more

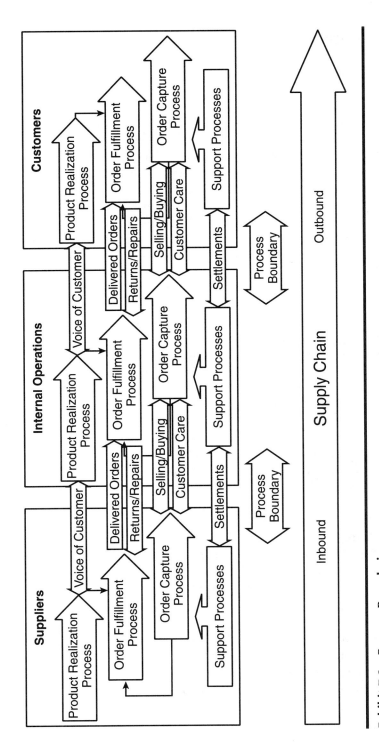

Exhibit 7.3   Process Boundaries

companies move to using on-line exchanges, electronic procurement systems, and Web-enabled demand management systems, there will be a growing trend to "impose" business processes on suppliers. These impositions will take the form of either adapting to an entire business process or being able to support a process interface.

The requirement to adapt to an entire business process is happening just as fast as the requirement to support an interface. An example of an entire business process being imposed by a customer on a supplier is Cisco Systems using it's own systems to plan and schedule the production of many of its contract suppliers. This type of relationship was reached after much negotiation and the building of a strong relationship. It also represents how a strategy for contract manufacturing can evolve to the point that it imposes a unique business process on one of its suppliers.

EDI is an example of supporting an interface. Because it has taken many years for industry to become comfortable with EDI, it has not become the predominant interface between buyer and seller on a business-to-business basis. The Internet-sired child of EDI is the electronic procurement system (EPS). An EPS incorporates not only order entry capabilities, but also on-line catalogs, contract compliance, and financial settlement functionality.

Process boundaries are growing in importance as industries and groups of companies in supply chains begin to collaborate on demand management, inventory management, and new product development. This is leading to a trend to have more real-time control using Internet and computer technology. As a result, if one company is negotiating a large, long-term agreement on a certain product line, it will likely also be negotiating a set of "external" business processes that will determine how each company in the supply chain will operate with each other. These external processes can be implemented via the customer's computer system, via the supplier's computer system, via an exchange provided by a third party, or via an entirely new web-based solution created specifically for this purpose. These new external processes must operate successfully with the internal processes already in place in the company's operations. These interfaces are critical. There is an entirely new software industry evolving to address these needs, and this software must be considered a member of the solution set.

The internal processes for which there will likely be boundary interfaces with external processes from customers and/or suppliers include the product realization process (PRP), order capture process (OCP), order fulfillment process (OFP), and the transaction settlement process (TSP).

The PRP is the process that takes new ideas and customer needs and converts them into manufacturable products. The internal process must be inherently cross-functional, data intensive, and parallel in nature. The external processes that the PRP must connect with include the customer's OFP, the customer's PRP, the supplier's PRP, and the supplier's OCP.

The OCP is the process that communicates with customers for the purpose of selling a product or service. The internal process must be inherently flexible but focused on satisfying customers needs. The external processes that OCP must connect with include the customer's OFP, the customer's PRP, and the customer's EPS.

The OFP is the process that converts a customer order into a delivered solution. The internal process must be inherently cross-functional, data intensive, and parallel in nature. The external processes that the OFP must connect with include the customer's EPS, the customer's PRP, the supplier's OFP, and the supplier's OCP.

The TSP is the process that converts receivables and payables into settled accounts. The internal process must be inherently cross-functional, data intensive, and parallel in nature. The external processes that the TSP must connect with include the customer's TSP, the customer's OCP, the supplier's TSP, and the supplier's TSP.

Examples of external process boundaries imposed by customers include interfaces between the:

- Materials management system of the customer and the production scheduling system of the supplier to support rapidly changing order and shipping schedules
- PRP of the customer and the PRP of the supplier to support cross-functional supply chain product development projects
- Materials management system of the customer and the order status system of the supplier to determine the status of existing orders

Examples of external process boundaries imposed by suppliers include interfaces between the:

- Materials management system of the supplier with the production scheduling system of the customer to support rapidly changing order and shipping schedules
- PRP of the supplier with the PRP of the customer to support cross-functional supply chain product development projects.
- OCP of the supplier with the purchasing system of the supplier to assist in the selection of product and making orders.

There are two key questions related to boundary definitions: What flows across the boundary? and What is the format of boundary flow? The information that flows across a boundary can be anything from online product catalogs, to demand information, legal documentation, intellectual property, messages, and database files. The format can be standard or custom. The issue of security must also be addressed, so now encryption technology becomes a major requirement for boundary flows.

In summary, a process boundary vision should include definition of:

- high-level process boundaries
- high-level information, work, and material that should cross the boundaries
- high-level ownership responsibility for the process boundaries

Control Vision

As discussed thus far, it is clear that e-factory business processes must be fast, flexible, and under control. The control factor is important because it is the secret to quality and cost-effectiveness. If a business process is not under control, there will likely be wide variability that leads to poor quality and ultimately higher costs. The cost factor is exacerbated by the fact that variability leads to inefficient use of resources.

There are two key parameters in process control: observability and controllability. Observability refers to the ability to observe, measure, and collect data at control points in a process that will provide information on the status of the process. Only data that is observable can be useful in calculating the control parameters. For example, if a manufacturing process depends on the temperature of the air surrounding a machine, then a parameter that should be measured is air temperature. If the air temperature is "observable," (i.e., measurable), then it is possible to perform calculations that can be used by an air-conditioning system to raise or lower the air temperature to the required level. Controllability refers to the ability to maintain certain parameters within a tolerance range. Continuing with the analogy, if there is an air-conditioning system that can be used to heat or cool the air to within the required range, then the air temperature is controllable. However, if there is no air-conditioning system, then even if it is possible to measure the temperature, it is still not controllable.

The controls vision should include a high-level definition of the:

1. Control hierarchy
2. Key control points
3. Key control parameters

Definition of the control hierarchy includes definition of each control level and how many control levels are required. The control points are those physical or software sites where control can be exercised. The control parameters are the values that have to be maintained within defined tolerances.

Technology Vision

The technology vision is one of the most important aspects of the visioning process. While it is neither likely nor necessary that a final technology solution set can be "visioned," there is a basic set of concepts that can be identified and accepted during a visioning process. The elements of a technology vision include a high-level definition of the:

1. Degree of interdependence of process and technology
2. Role of data and knowledge bases
3. Role of hardware and software
4. Performance goals and parameters

A definition of the degree of process and technology interdependence is required because one of the fundamental concepts of the e-factory is that technology accelerates and automates critical elements of business processes. The technology vision should address at a high level how extensively technology should be used for both the horizontal and vertical processes; the number of levels in the control hierarchy and how tightly coupled those levels should be; and how much automation should employed. Another key question is whether technology or people will be in a supporting role or a critical role in each of the process activities.

The role of databases and knowledge bases is an extremely important issue for virtually every aspect of the e-factory. All of the solutions set components discussed herein are based on using, adding to, or modifying databases. The databases contain product information, order information, price information, customer information, sensor information, and process information, and become knowledge bases when there is information included that describes decision rules, relationships, and conditional events. Most automation technologies available today depend on real-time access to specialized databases. Not only is it important to define a vision for the role of data/knowledge bases, but it is also important to determine whether the data/knowledge bases exist and how they are to be built if they do not already exist.

An important but necessarily critical issue that should be addressed in a technology vision is the high-level role that both hardware and software play in the e-factory. For many of the real-time performance requirements, hardware performance will be a gating performance factor for the solution set. Much of the factory floor automation and all of the intelligent process equipment will have embedded microprocessor technology. If a real-time network is required, there will be hardware implications. While the final sizing of solution-set hardware and software will be done as part of the e-factory roadmap, there will be a need to have a high-level vision of how much hardware power will be required to execute the other elements of the vision.

Probably the most important element of the technology vision is the set of performance goals and parameters that will be used to measure the success of the e-factory solution set. The best-practice set of performance parameters includes cycle time (e.g., customer response, order fulfillment, Web site response, etc.), cost per cycle, cost per unit, and first-pass yield.

Change Management Vision

Change management is the process of managing an organization's ability to deal with and implement change. Implementing an e-factory usually involves significant change for the workforce. Changes typically come in the form of training to perform new jobs, training to use new technology, working in a different supervisory environment, all being measured by different types of performance parameters and being compensated by different types of reward systems.

The key elements to a change management vision should include:

- Feedback and improvement methodology
- Frequency and nature of communication
- Training and recycling skill sets

The most important change management component is how feedback from people in the e-factory business processes will be collected and used to improve the effectiveness of the processes. This is a non-trivial concern because, with the large investment that most companies make to implement an e-factory, there should be a mechanism in place to ensure that opportunities or necessities for improvement are pursued. Typically, some type of continuous improvement program is needed. These programs usually include frequent (daily, weekly) meetings of process teams, training in problem-solving and idea-generation tools for the teams to use, and visible performance measurement systems so that people can see when their improvement ideas are working.

A common denominator for all forms of change management is communication. Such communication should include information about plans, implementation programs, performance, and rewards. The more information there is available, the less chance there is of misunderstandings being created by speculation about the unknown. Good communication is synonymous with successful change management practices. The change management vision should include a vision of how much change management will be required and how it will be implemented.

A key part of change management implementation is how much skill recycling, training, and new recruiting will be required to support and operate the e-factory environment. The extent of change in skill sets will drive how much new training and new recruiting will be necessary.

Top-Down versus Bottom-Up Visioning

The whole purpose of visioning is to establish the basic of a framework of how the e-factory should work, or at least what it should accomplish. The vision in reality becomes an initial condition of the goal line. As the roadmap development methodology unfolds, this initial condition may evolve until the vision changes to meet the defined solution set.

The visioning process can begin from the high-level perspective first, with deeper levels of detail developed as necessary. The visioning process might also begin from the lowest-level perspective and build up to a high-level strategy. Both of these directions — top-down and bottom-up — are reasonable and viable. In fact, an iterative process using both the top-down and bottom-up approaches is often the best because it allows the essence of a vision to be tested from both perspectives.

Visioning from the top down begins with developing a plant-level vision and then drills down through each control level. The top-down approach is driven by go-to-market requirements. It stimulates process strategy development.

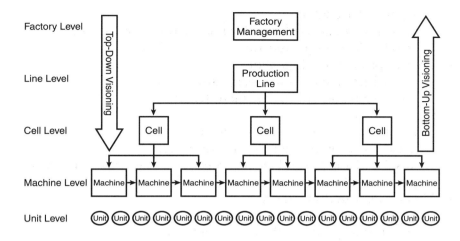

Exhibit 7.4   Iterative Visioning

Visioning from the bottom up begins with process scenarios that define the key activities that must be performed to achieve the business process goals (see Exhibit 7.4).

Using the enterprise supply chain model presented in Chapter 1, it is clear that there are three potential sources of requirements for the e-factory: (1) the set of customers being served by the enterprise and its supply chain; (2) the set of suppliers that serves the enterprise with materials, products, and services; and (3) the set of internal business processes operating within the enterprise. To define the scenarios at the process level, it is important to use the requirements dictated by all three sources of requirements for the e-factory. One of the benefits of this visioning activity is to not only generate a vision for the e-factory roadmap, but to also stimulate the capture of requirements from the three sources.

Customer-Driven Requirements in the Supply Chain

The requirements that are the most important to the creation of an e-factory vision are those created by the customer community. Because the entire purpose of an enterprise and the business processes that compose it is to profitably satisfy customer needs, these requirements must be identified as early in the process as possible. Exhibit 7.5 illustrates a typical set of process relationships that generates the customer-driven requirements.

A typical set of customer-driven requirements includes:

1. Standard order content
    a. Standard product configuration
    b. Standard delivery requirements
    c. Standard payment terms

2. Custom order content
   a. Special first-, second-, or third-tier supplier instructions
   b. Special delivery instructions
3. Special product configuration
   a. Special production processing instructions
   b. Special pricing and/or payment terms
4. Order status
   a. Location of order in supply chain
   b. Location of order with production process
   c. Results of order production process (quality, cost, uniqueness, key parameters)
5. Vendor managed inventory
   a. Customer needs identified by supplier
   b. Supplier responds with replenishment
   c. Order processing initiated by supplier

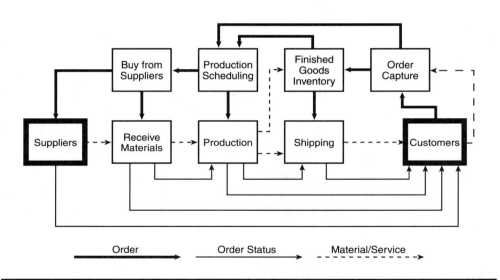

Exhibit 7.5   Customer-Driven Requirements

Supplier-Driven Requirements in the Supply Chain

The set of requirements that are quite often the most forgotten during the planning for the creation of an e-factory vision are those created by the supplier community. Because the performance efficiency of an enterprise and its business processeses depend on how effectively the suppliers operate, the supplier requirements must be identified as early in the process as possible. Exhibit 7.6 illustrates a typical set of process relationships that generates the supplier-driven requirements.

Examples of supplier-driven requirements in the supply chain include:

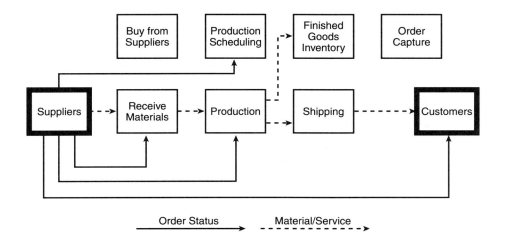

Exhibit 7.6 Supplier-Driven Requirements

1. Schedule coordination with purchasing
   a. Status of order
   b. Coordination of schedule
2. Schedule coordination with production line
   a. Status of order
   b. Coordination of schedule
3. Schedule coordination with customer
   a. Coordination of delivery schedule of customer specified materials/ services
   b. Planning
4. Receive feedback on performance
   a. Delivery
   b. Quality

Internal Process-Driven Requirements in the Supply Chain

The requirements that are usually the only ones actually considered — and usually not very thoroughly — are those created by the process needs internal to the enterprise. Because the operation of the internal processes is the performance that customers measure when they determine who they want to do business with, the internal process-driven requirements must be identified as early in the process as possible. Exhibit 7.7 illustrates a typical set of process relationships that generates the supplier-driven requirements.

The production-driven requirements include:

1. Build to forecast
   a. Provide capacity information
   b. Coordinate schedules with other plants
   c. Ensure that proper features and machine settings are used

2. Build to order
   a. Confirm order start
   b. Provide order status
   c. Provide order results
   d. Coordinate requirements with suppliers
   e. Ensure that proper features and machine settings are used
3. Production line optimization
   a. Choose optimal mix for production period
   b. Coordinate schedules with other plants
   c. Coordinate requirements with suppliers
4. Order fulfillment optimization
   a. Choose between stock versus production
   b. Coordinate schedules with other plants
5. Process instructions
   a. Recipes
   b. Parts
   c. Assembly instructions
   d. Quality requirements
6. Machine settings
   a. Ensure match with process instructions
   b. Match with operator instructions
7. Machine results communicated to other machines
   a. Feedback on quality issues
   b. Feed forward on work instructions
8. Materials properties and status reported to machines
   a. Material may carry identity, location, operating instructions
   b. For both operator and machines
9. Cell controllers rescheduling production
   a. Cell controllers used for machine clusters or focused line
   b. Other levels of control communication
10. Maintenance requirements planned
    a. Machine usage feedback to maintenance management system
    b. Real-time maintenance or calibration settings

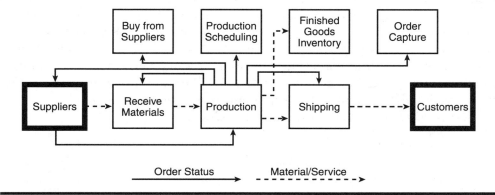

Exhibit 7.7   Internally-Driven Requirements

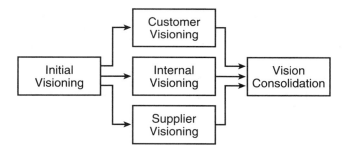

Exhibit 7.8   Vision Creation Process

Creating the Vision

Creating the vision of an e-factory is an important first step in building a roadmap for specifying and implementing the e-factory. It is important because it stimulates and initiates the effort necessary for capturing roadmap requirements that emanate from the three key supply chain participants: customers, suppliers, and internal operations.

The process for creating the vision is not complicated but does require planning. The flowchart in Exhibit 7.8 outlines a simple process structure for creating an e-factory vision.

The first step is to organize and launch a visioning project. This step consists of forming a visioning team, selecting members for the team, creating a team goal statement, and beginning the initial visioning activities. The elapsed time for an initial visioning effort should take between two and four weeks. Any longer and the initial effort runs the risk of collapsing due to neglect.

Because of the three different sources of requirements for the e-factory, there should be three distinct visioning efforts or projects spawned by the initial visioning session. Each effort should be focused on one of the three requirements source groups. The reason that a separate visioning effort is required for each is to ensure that an intense focus is placed on capturing the requirements specific to that source.

Each of the three requirements source vision efforts should include getting direct input from representatives of the sources. The customer visioning effort should include the involvement of customers. Likewise, staff from suppliers should be involved in the supplier visioning effort. The internal source should be involved in the internal visioning session as well. Although it may seem obvious that staff operating in the internal processes should be involved in the internal visioning effort, they may be the most difficult to schedule or enlist because of their daily duties. If the e-factory is a greenfield facility, the internal staff may not have all been hired at the time the visioning must occur.

The final step is the consolidation of all the visioning developed to that point. If all the individual visioning efforts result in similar e-factory visions, then the consolidation can be a very simple process. If the individual visions have serious conflicts in structure or focus, then there is a need to spend more time understanding the conflicts and resolving them. Resolution of e-factory vision conflicts may require compromise, consolidation, or rework of the individual vision elements to achieve an integrated vision. If the conflicts in the e-factory vision are so serious that they defy compromise or consolidation, then there might be serious issues about the overall strategy for the e-factory in that particular situation. In this case, a review of the strategy should be done before going any further with an e-factory strategy.

Summary

The visioning process for the e-factory is an important step in creating the roadmap. Creating a vision is a way to get a roadmap team working together. It is a way to get management and operations to clarify what their goals and objectives are for the e-factory. More importantly, it is also a way to get started on capturing the requirements from the three key sources that have to be satisfied by the e-factory. The three areas are suppliers, internal operations, and most importantly customers. By collecting the requirements, the vision that gets created is not solely the result of the brainstorming and discussions of people internal to manufacturing business processes.

Chapter 8

An e-Factory Roadmap Methodology

Purpose of This Chapter

Because the concept of an e-factory is new and rapidly evolving, a methodology for creating a roadmap of how to design and select the best technology and processes for an e-factory can be an important tool. There are many factors to consider, many new requirements, and many alternative solutions to evaluate when planning for an e-factory. By having a methodology that provides for the rapid structuring of issues and that forces the evaluation of key factors, the quality of the decisions being made is improved. The purpose of this chapter is to describe a methodology for creating a roadmap for defining and implementing an e-factory.

Introduction to the Methodology

The methodology presented in the following is designed to force and facilitate the creation of an e-factory roadmap. This methodology is an organized approach to performing the necessary analysis and making the necessary decisions that create a roadmap for an e-factory. One of the key components of this methodology is scenario analysis. Another key component is object-oriented database technology. Both of these components are used to organize and manipulate the data necessary for the decision-making process.

A flowchart of the methodology is shown in Exhibit 8.1. The methodology starts with the formation of a roadmap team. The team should be cross-functional and represent internal operations, customers, and suppliers. The second step is the gathering of requirements (see Chapter 7) generated by customers, by internal processes, and by suppliers. The third step is the

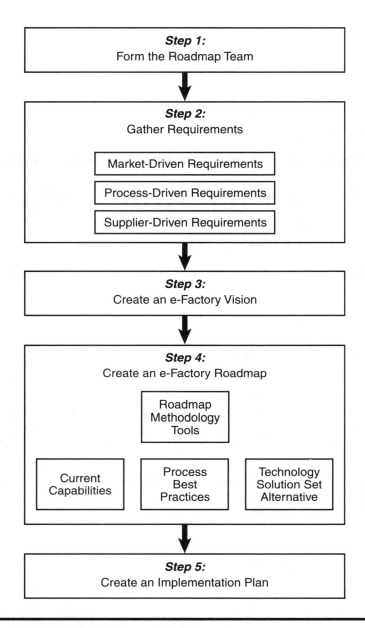

Exhibit 8.1   Roadmap Methodology Flowchart

creation of a vision of the e-factory. The vision will assist in focusing the roadmapping effort. The fourth step is the creation of a roadmap. In this step, one or more scenarios are generated to describe the flow of work, materials, and information through the e-factory.

There should be enough scenarios generated to cover all the key business processes and key variations in how business is done. When finished, the scenarios are used to create and map information about the solution sets and about the process alternatives into the data tables that together form the decision tool. The tool provides a place to summarize the analysis done by the roadmap team as it works through the issues and creates the roadmap. The tool can be implemented in either simple spreadsheet technology or can be built in any of the popular database technologies available on the market today (e.g., Microsoft Access or Oracle8). The tool used for this book was created in Microsoft Access.

The key challenges in developing a roadmap for an e-factory are twofold: (1) how to collect, organize, and use all the information that should be available from all the supply chain sources; and (2) how to perform the analysis and make all the decisions about the structure and content of an e-factory roadmap.

This methodology uses scenarios to generate (1) key relationships between requirements, processes, and technology, (2) decision tree-like results for making the best decisions, (3) requirements from customers, suppliers, and internal process needs, and (4) support for an iterative bottom-up and top-down approach. After generating the scenarios, the database tables can be used to assist in the selection of the best combination of information technology solution set components and business process designs. The data tables used as part of the methodology are as follows.

1. *Scenario definition table:* describes at a high level the purpose of and differences between the various scenarios that have been or should be generated to describe the basic nature of the e-factory in question
2. *Scenario generation table:* documentation of the specific activities executed, objects involved, and requirements driving each scenario generated; also documented are the key parametric measures (e.g., cycle time, cost) of each scenario
3. *Activity-based cost generation table:* activity-based cost components for each activity in each scenario
4. *Requirements definition table:* list and definition of requirements coming from customers, suppliers, or internal process sources
5. *Activity definition table:* list and definition of activities in each of the scenarios
6. *Object definition table:* list and definition of objects involved in all the scenarios
7. *Solution component capabilities table:* for each solution set component, a list of capabilities that represents a mapping from the general capabilities of solution set component into the scenario requirements
8. *Solution set mapping table:* matrix used to map the solution set candidates into applications for each scenario
9. *Requirements value definition table:* definition of the requirements value criteria
10. *Cost value definition table:* list and definition of the cost value criteria
11. *Time value definition table:* list and definition of the time value criteria
12. *Solution set evaluation table:* summarizes the value of how effectively each solution set candidate satisfies each requirement, cost, and cycle time goal

13. *Implementation requirements table:* list and definition of the fundamental steps in implementing an e-factory solution set
14. *Solution set roadmap table:* identifies the best steps through the implementation process in terms of sequence and implementation requirements

A simple diagram that shows the relationships of the tools is given in Exhibit 8.2. A brief introduction to each table is presented in the following chapter sections.

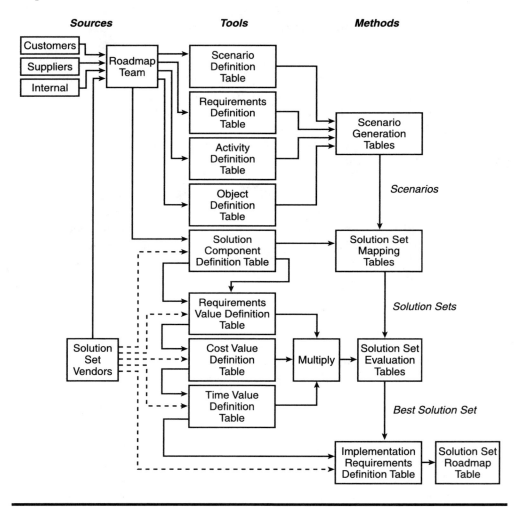

Exhibit 8.2   Method Tool Flowchart

Introduction to Scenarios

The use of scenarios is being introduced as a key part of the roadmapping methodology because it provides a way to convert descriptions of how a business process is supposed to work into requirements for workflow, material flow, and information flow. To understand how a scenario might be created to describe a business process, consider a situation in which a New York Yankee fan is asked to describe to a new arrival from a distant planet how a baseball game works. The fan might say something similar to:

- A person, called the pitcher, throws a spherical mass, called a baseball, to a person, called the catcher.
- A person, call a hitter, tries to strike the baseball with a cylindrical shaped object called a bat.
- If the hitter strikes the baseball in a direction that falls into a territory designated fair, then the hitter must run to a station called first base.
- And so on…

It is easy to see that the entire process called a baseball game could be described by a (very long) series of statements that could cover every eventuality. While this list of statements would be tedious for the person making the explanation, as well as for the alien, it would provide a thorough description of what might happen during a game. It would also be extremely useful for capturing all the rules, relationships, and interactions that would have to be programmed if a video game that simulates a baseball game was going to be designed.

Scenarios have been used since the early days of systems analysis as a way to describe how a system or process might operate given a set of initial boundary conditions. Scenarios can be described with a variety of techniques that include flowcharts, process maps, data flow diagrams, and activity tables. Graphical devices such as flowcharts and process maps have long been the backbone of business process improvement and systems modeling techniques because the process of creating them is visual and is amenable to generation by a team of people in a war room-like setting. Typically, the process map documents a steady-state model of the resources and sequence of activities through which work and data flow. These work and data flows are best described as a set or list of activities that reflect the interaction between the participants or objects in the process. Building process models using object technology concepts adds meaning and creates a vital link to the design and deployment of information technology by identifying the transactions and relationships between resources within the process.

In recent years, object-oriented technology has evolved as a technique for designing systems because it provides for a richer and more rational approach to defining how different types of inanimate or animate entities interact with each other. Transitioning to the use of object technology for process modeling requires the introduction of a few simple concepts about entities and relationships, and the flow of interactions between the entities through the relationships.

For example, an object is a resource (which can be human, machine, or data) that can contain information or behavior. These objects have relationships and experience transactions with each other. Characterizing the information attached to objects and defining the relationships and transactions between them are key aspects of how object technology can be used to enhance business process modeling. The technique that has evolved as the best way to create the initial definition of objects and relationships is the scenario.

The typical business process map tends to be a composite of a variety of paths through the map that depend on initial conditions and the decisions that could be made at key decision points within the process. During the course of building a business process map, especially when using a routing-by-walking-around method, the alternative paths (or process scenarios) can be identified. A scenario map can then be created that defines an alternative path of activities and the transactions between the objects during this path. The scenario map thus becomes a structure for specifying the objects involved, defining the inputs required and outputs expected at each step, estimating resource consumption, and specifying data flow between objects.

A convenient way to represent information generated by the creation of a scenario map is through the use of a scenario matrix. A scenario matrix consists of the sequence of activities in the scenario path; the array of objects involved in the scenario; and for each activity, the transaction between the involved objects. The transaction can be between two objects or within one object. Transactions can involve the flow of work, material, data, or responsibility. This matrix represents a variety of information structures. For example, the columns of information resource objects with the included data transactions contain the information that would be found in a data flow diagram. The columns of human resource objects with the included transactions contain the information that would be found in the typical process map. The partial example below represents a scenario matrix for a product realization business process.

The use of object techniques for process modeling suggests the use of object-oriented case tools for the creation, management, and manipulation of the scenario maps and matrices. One of the benefits of using these enabling tools is that they allow the generation of data flow diagrams, process maps, and process simulation using the same object-oriented scenario database. They also allow the development of the information system design concurrently with the process design. These sophisticated tools increase the effectiveness of the process model and speed up the implementation of both systems and workflow.

One of the key features of these packages is the ability to decompose a process from high-level phases into more detailed levels and finally into activities. This decomposition allows the process designer to first develop a process vision and then focus on defining the details for specific objects within subprocesses. The process decomposition feature allows for gradual reengineering and improvement over time within the context of an overall vision.

Flexibility and speed are the promises of object-oriented business process modeling. Flexibility in that it allows the process designer to easily consider "what-if" process alternatives and to reuse or modify process models. Speed in that it allows for the rapid configuration of system solutions that support the process solution.

The use of object-oriented business process models will expand in the future because of the object-oriented trend in enterprise and application systems. As emerging software products provide for implementation configuration techniques using object technology, designing a business process using object technology simultaneously produces the system design. This will stimulate the introduction of best-of-breed solutions that also employ best practice process solutions as a means of achieving successful customized solutions while increasing the speed of change.

The scenario tool is a great way to force the definition and structure of a multitude of information that represent the requirements driven by customers, internal production processes, and suppliers. The most basic information is the set or list of activities in a process. This list of activities describes the flow of material, information, work, and responsibility through a business process. The next most important information is the set of objects that must operate together in a business process. This set of objects includes humans, machines, and information technology objects such as databases, software applications, and network resources. The third important set of information is the set of relationships of the entities. Relationships include which entities are involved in each activity, what the entities are doing to each other or the material, and what requirement they are satisfying.

The beauty of scenarios is that they can be built for different business processes, for different products within a business process, and for different assumptions about the boundary conditions. Thus, as many scenarios can be built as need to be built to describe the reality of the situations involved in a business process. Because the requirements that are driving the e-factory are so new and rapidly emerging, many of the situations that might exist must be hypothesized because there may not be any previous experience to act as a guide. Thus, the process of creating scenarios forces an engineering team to deal with the details of a potential reality.

Because the scenario approach is object technology based, it can be done in graduated levels of detail. For example, a process workflow can be defined for a high-level set of objects first using workflow steps. To go to a lower level of detail, each workflow step can be broken down into activities and then subactivities at an even lower level. As each lower level of detail is defined, the new objects may need to be defined and included in the object set.

It is helpful to use an example that is as simple but as meaningful as possible to illustrate the concepts and the tools described in this chapter. A process flowchart for a simple manufacturing line is shown in Exhibit 8.3. This example is used throughout this chapter. The flowchart captures the essence of how a customer order is fulfilled through the conversion of material purchased from a supplier into a finished product. This is a very high-level representation of how a manufacturing company operates. All of the activities and all of the human, machine, and data resources have been lumped into high-level process steps.

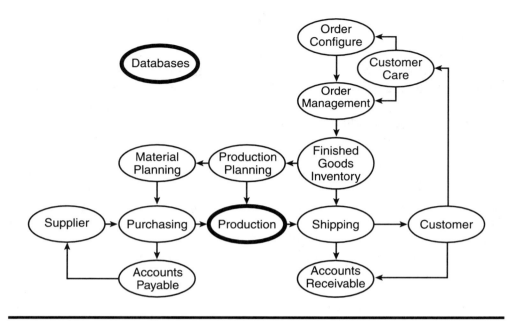

Exhibit 8.3   High-Level e-Factory Process Map

When the concept of objects is introduced and taken to a greater level of detail, the process flowchart takes on a slightly different character, as indicated in Exhibit 8.4. In this diagram, "objects" such as databases, cell controllers, and people have been introduced to better describe the actual flow of work, information, and material between humans, machines, and databases. By expanding the scenario to include animate and inanimate objects, not only can the reality of the e-factory be better described, but also the trade-off between automation technology and manual processes can be better evaluated. Solution set alternatives can also be compared and evaluated using the same scenario.

If it is assumed that the simple example in Exhibit 8.4 is for a company that manufactures a simple product such as a toy firetruck, the production process could be segmented into three key process steps. The first step is the fabrication of the metal parts for the truck. To keep the example simple, assume that all the parts can be stamped from sheet metal. The second key step is the coating of the parts with paint. The third step is the assembly of the parts into a finished truck. The truck can come in different colors and with different accessories or configurations of features such as extended cabs, additional exterior lamps, and different truck bed configurations.

Using the diagram in Exhibit 8.4, several different scenarios can be generated, depending on many assumptions about how customers are to be served, how suppliers are to be employed, and how the operations in production and customer service are to be executed. For example, one scenario could be based on the assumption that all products are designed to be built-for-stock, are standard products listed in a published catalog, and are sold by a

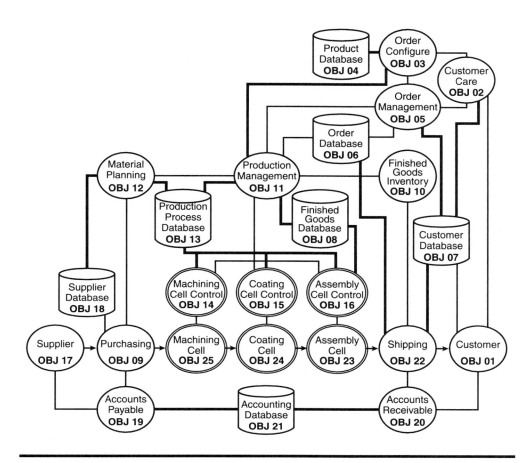

Exhibit 8.4   High-Level e-Factory Scenario Map

field salesforce. Another scenario is that rather than being sold by a field salesforce, product catalogs are on-line on the company Web site, and customer purchase orders are received on-line. A third scenario is that products are designed to be built-to-order and sold over the Internet. In the third scenario, there would be very little inventory. The production processes and the supplier responses must be fast compared to the order fulfillment cycle time promised to the customer. These three scenarios are different in many ways. They could represent three different business models that might require a different company for each scenario. However, many manufacturing companies today are facing the situation where they have to support all three scenarios and probably many other variations. This need for diversity in business scenarios is being created by the need to support a diversity of marketing channels. With the transition to a computer-based e-business environment, the need to expand and diversify supply chain scenarios is essential.

In the following paragraphs, examples of the various tables employed in the roadmapping methodology use the simple example in Exhibit 8.4 as the basis for illustrations.

Scenario Definition Table

Most manufacturing environments have more than one operating mode or more than one set of conditions under which it operates; that is, most plants are not built, set up once, and then just turned on to run with no variation in product, lot size, or quality. Therefore, more than one scenario will likely be required, as well as a high-level description of each. The Scenario Summary Table is a simple high-level documentation of the purpose and differences of the various scenarios that have been or should be generated to describe the basic nature of the e-factory in question. A simple example of a Scenario Summary Table is illustrated in Exhibit 8.5.

Exhibit 8.5   Scenario Summary Table Example

ScenarioID	Scenario Description
1a	Build-to-stock, catalog products, published catalogs, field sales transactions, blanket orders for suppliers, production planning based on long-term forecasting
1b	Build-to-stock, catalog products, on-line catalogs, field sales promotion, blanket orders for suppliers, production planning based on long-term forecasting
1c	Build-to-stock, catalog products, on-line catalogs, field sales promotion, supplier-managed inventory, and production planning based on long-term forecasting
2a	Build-to-order, catalog products, on-line catalogs, field sales promotion, blanket orders for suppliers, production planning based on short-term order trends
2b	Build-to-order, catalog products, on-line catalogs, field sales promotion, blanket orders for suppliers, production planning based on short-term order trends
2c	Build-to-order, catalog products, on-line catalogs, field sales promotion, supplier-managed inventory, and production planning based on short-term order trends

In this example, there are two major types of scenarios. One is build-to-stock, and the other is build-to-order. The variations in the two major types illustrated differ in the degree of automation and degree of involvement by suppliers in operations. These scenarios could represent different product lines in one company or alternative strategies for one product line. If the situation is the latter, then the roadmap methodology can be used to evaluate strategy as well as solution set.

Scenario Generation Table

An important step in the roadmap methodology is the generation of scenarios that describe the business processes in the e-factory. Generating the scenario

Scenario ID	▼
Activity No.	▼
Business Process	▼
Activity	▼
Performing OBJ	▼
Source OBJ	▼
Receiving OBJ	▼
Requirement	▼
Required Cycle Time (min)	▼

Exhibit 8.6   Generic Scenario Generation Form

requires articulating a description of the story of how a business process works, as the lighthearted baseball example above illustrates. Generating a scenario requires that each activity be described and each object involved in the activity be identified. A generic form that can be used to capture such information and have it stored by a computer-based scenario generator tool is shown in Exhibit 8.6. A sample of the database table that this form generates is shown in Exhibit 8.7. This table is, in part, a relational database. The vertical axis represents the list of activities. The horizontal axis includes information about activities, objects that are involved in each activity, and the properties for each of the objects.

The table in Exhibit 8.7 consists of several columns and rows. Each row in the table is information that describes an activity and other relevant information. Each column represents information necessary to the description of the scenario. The first column is a number used to identify the scenario. As discussed above, there often will need to be several scenarios generated to reflect the reality of the situation facing an e-factory design team. The second column is the number of the activity in a scenario. The total number of activities for each scenario could range from a few dozen for a high-level description, to hundreds or thousands for a more detailed description of the total business process. Of course, the greater the level of detail, the greater the number of activities in a scenario. The greater the number of activities, the more time-consuming is the creation of the scenario and the more difficult is the handling of the scenarios. The trade-off between the level of detail and the speed of creating and using the scenarios depends on the ultimate purpose. For building an e-factory roadmap, staying at the higher level of detail is often adequate. For building a detailed implementation specification for software, job descriptions, control algorithms, etc., more detail is necessary. The Scenario Generation Table can ultimately act as the documentation method for the final implementation of an e-factory process.

Exhibit 8.7 Sample Scenario Generation Table

Col.1 Scenario ID	Col.2 Act. No.	Col.3 Business Process	Col.4 Activity Description	Col.5 Source OBJ	Col.6 Performing OBJ	Col.7 Receiving OBJ	Col.8 Requirement	Col.9 Req. Cycle Time (min)
1b	1	CRM	Customer contacts company via Internet	1	1	2	Answer customer questions about products	0.25
1b	2	CRM	Customer requests information about product features and pricing	1	1	2	Answer customer questions about products	0.25
1b	3	CRM	Customer care Web pages calls for product configurator program	2	1	3	Answer customer questions about products	0.01
1b	4	CRM	Order configurator program takes question and queries product database	3	2	4	Answer customer questions about products	0.01
1b	5	CRM	Product database responds with answer to customer question	4	4	3	Answer customer questions about products	0.25

The third column is the name of the business process. This information can be useful for management purposes and for deciding on the priority of implementations. Typically, the business processes involved include order capture, order management, customer relationship management, supply chain management, and order fulfillment. The fourth column is the description of the activity being executed. Typical activities include movement of material, movement of data, movement of instruction, turning machines on or off, sending a status report or signal to a machine or person, or collecting information. For example, a material movement activity could be machine-to-machine, within-a-machine, facility-location-to-machine, or person-to-machine. An information movement activity could be person-to-IT-resource, person-to-machine, machine-to-IT-resource, material-to-person, material-to-IT-resource, or material-to-machine movement. A machine operation activity could be the beginning-of-operation, completion-of-an operation, or intermediate-step-of-operation.

The fifth, sixth, and seventh columns identify which objects are involved in the activity. There are three types of involvement an object can have for each activity. An object can be a performing object in that it is the object that performs the activity; a receiving object in that it receives the action, information, or material that is transferred during the activity; or a source object in that it is the source of the information or material that is transferred during the activity. There can be three different objects serving each of these purposes, or one object can serve all three. If the object performs the activity and does not interact with any other objects during the performance, then the same object number will be in all three columns.

The eighth column is the description of the requirement being satisfied by the activity. The purpose of this column is to ensure that all the requirements that have been identified as being necessary for market, process, or supplier reasons are being satisfied by the actual business process as defined by the scenarios. The degree of satisfaction of these requirements and the cost of satisfying these requirements will be useful in the final evaluations of the solution set alternatives.

The ninth column is the definition of the average or typical cycle time of the activity. Cycle time refers to the elapsed time required to perform the activity. This time factor is important because it will contribute to the overall cycle time of the entire process and it will be a cost driver. This is a key metric for determining which solution sets and which scenarios will be the most effective.

Although the Scenario Generation Table is designed to ensure that all the appropriate information is collected, scenarios are best built by the people that are either already working within the process or those who have been chosen to design the process. To build a scenario, a cross-functional team of people focused on a specific process should be formed and launched.

Referring back to the simple e-factory example in Exhibit 8.4, a portion of a sample Scenario Generation Table is given in Exhibit 8.7. This simple example actually has 68 activities in scenario 1b alone.

Activity-based Cost Generation Table

Shown in Exhibit 8.8 is a table of the activity-based cost drivers that can be used to evaluate the cost of each solution set candidate for each scenario. The remaining columns in the table represent the process metrics that will be used to calculate the value of the different solution set alternatives, but will also provide a way to measure the effectiveness of a particular scenario implementation of human, machine, and data resources.

Column 1 is the scenario identification. Column 2 is the identification of the activity. Column 3 is the identification of the solution set for which the costs are being calculated. Column 4 is the identification number of the solution component that can be applied to this activity. Column 5 is the identification of the object involved. If more than one object is involved (e.g., a source, performing, and receiving object), then each object should have a separate line. Column 6 is the definition of the amount of labor that may be consumed during the activity. If the object involved in the activity is a human resource, then labor will be consumed. The amount of labor is important because it will be converted to a cost using the average cost rate for the labor category used for the activity. If no labor is used in the activity, then the entry for that activity would be zero. Column 7 is the amount of machine time consumed during the activity. If the objects involved in the activity is a machine resource, then machine time with all the accompanying wear and supporting utilities will be consumed. The amount of machine time is important because it will be converted to a cost by one of the following columns. If no machine is used in the activity, then the entry for that activity would be zero. Column 8 is the amount of other costs consumed during the activity that is not associated with the labor and machinery already considered by the columns above.

Column 9 is the total unit level costs. This is the sum of all the costs associated with the labor, machine, and other costs created by the three previous columns. The unit level cost is one of the four levels of costs that compose an activity-based cost structure. The other three levels of cost are batch, sustaining, and facility. Unit-level costs include all costs associated with the production and process of one unit of output for the business process scenario being generated. Column 10 is the total batch-level cost. The batch-level costs include all costs associated with the setup and processing of batches in the business process scenario being generated. Column 11 is the total sustaining-level cost. The sustaining-level costs include all costs associated with the maintenance, continuing engineering, and upgrade work associated with the business process scenario being generated. Column 12 is the total facility-level cost. The facility-level cost includes all costs associated with the operation of the facility occupied during the business process scenario being generated. Column 13 is the total of all cost levels.

Requirements Definition Table

The Requirements Definition Table, an example of which is shown in Exhibit 8.9, is a list of all the requirements that were identified during or prior

Exhibit 8.8 Activity-Based Cost Generation Table

Col.1	Col.2	Col.3	Col.4	Col.5	Col.6	Col.7	Col.8	Col.9	Col.10	Col.11	Col.12	Col.13
Scenario ID	Activity No.	Solution Set ID	Solution ComponentID	Object ID	Labor Time (min)	Machine Time (min)	Other Costs	Total Unit Level Cost	Total Batch Level Cost	Total Sustain Level Cost	Total Facility Level Cost	Total Cost
1b	1	A	3	OBJ 02	0.0	0.01	$0.02	$0.04	$0.04	$0.03	$0.03	$0.14
1b	2	A	3	OBJ 02	0.0	0.01	$0.02	$0.04	$0.04	$0.03	$0.03	$0.14
1b	3	A	3	OBJ 02	0.0	0.01	$0.03	$0.06	$0.05	$0.05	$0.04	$0.21
1b	4	A	3	OBJ 02	0.0	0.01	$0.03	$0.06	$0.07	$0.07	$0.08	$0.28
1b	5	A	3	OBJ 02	0.0	0.01	$0.04	$0.06	$0.07	$0.07	$0.08	$0.28
1b	6	A	3	OBJ 02	0.0	0.01	$0.03	$0.05	$0.06	$0.06	$0.07	$0.23
1b	7	A	3	OBJ 02	0.0	0.01	$0.01	$0.04	$0.04	$0.05	$0.05	$0.19
1b	8	A	3	OBJ 02	0.0	0.01	$0.01	$0.04	$0.04	$0.05	$0.05	$0.19

Exhibit 8.9   Sample Requirements Definition Table

Req. Driver	Requirement	Total Activities
Internal	Electronically distribute production process instructions	4
	Execute customer order	4
	Execute customer settlement	2
	Execute order production	15
	Plan materials requirements in real-time	3
	Schedule order production in real-time	2
Market	Accept customer order	2
	Answer customer questions about orders	12
	Answer customer questions about products	7
	Provide customer with a quotation	5
Supplier	Execute materials purchase in real-time	8
	Execute materials shipment in real-time	3

to the scenario generation process. Requirements are generated by considering the needs of customers, suppliers, and internal processes. The scenarios should consist of activities that address all of the requirements from all three sources of needs. The first column in the table identifies which of the three sources or driving influences has created the requirement. The second column is the definition of the requirement. The third column is the number of activities in each scenario that is contributing to the satisfaction of the requirement.

There are at least two ways to generate the requirements. One approach is to take a cross-functional team consisting of representatives from all three communities of interest — customers, suppliers, and internal processes — and begin generating scenarios that reflect what the total team believes is necessary. As each activity in each scenario is defined, a corresponding requirement is defined. When the scenarios are finished, then the total set of requirements defined during this process can be compiled. This compiled list then reflects what the scenarios generated actually satisfy. This list can then be reviewed with a larger number of representatives from each group to stimulate discussion as to whether the compiled set of requirements adequately represents the true requirements. If new requirements are identified during this review, then either steps already defined may satisfy them or new steps will have to be created and inserted into the appropriate scenarios.

A second approach is to generate the requirements first, and then use the list of requirements to generate the scenarios. The requirements list can be generated by communicating directly with the communities through a variety of techniques. Customer requirements can be collected through surveys, focus groups, quality function deployment (QFD) sessions, and direct interviews. Supplier requirements can be collected through the same set of techniques. Internal process requirements are best defined using QFD sessions or direct interviews.

Activity Definition Table

The Activity Definition Table should be viewed as a simple compilation of all the activities defined during the scenario generation process. The purpose of this table is to serve primarily as a documentation of the specific activities contained within the scenarios. This table is useful because it serves as a record of the business process being updated or implemented, statement of the design objectives, and a model or template for future process changes.

The example in Exhibit 8.10 illustrates this simplicity because there are only two types of information: the definition of the activity and the scenario containing the activity. In fact, additional columns can be included to identify which levels, objects, and requirements are associated with the activities. This table also serves as a source of activities that might be used in other business processes. By dragging along the related objects and the solution sets, there is a potential for reusability of either the solution set application or the portions of the scenario using the activities.

Exhibit 8.10   Sample Activity Definition Table

Scenario ID	Activity
1b	Customer care Web page sends answer to customer
1b	Customer care Web page calls for product configurator program
1b	Customer contacts company via Internet
1b	Customer requests information about product features and pricing
1b	Customer sends payment to accounts receivable
1b	Order configurator program sends answer to customer care Web page
1b	Order configurator program takes question and queries product database
1b	Order management database notifies order management program of shipment

Object Definition Table

The Object Definition Table is the compiled list and description of all the objects involved in the scenarios. This list can include other information that is relevant to the objects, such as the type of object (human, data, or machine), the control level of the object, and the name of the object. An example of such a table is given in Exhibit 8.11. This table can also include other relevant information, such as the solution sets associated with each object.

Solution Component Capabilities Table

Because the e-factory solution set is likely to be a best-of-breed configuration, it will likely consist of a variety of software and hardware components that come from a variety of suppliers. Even if the solution set comes from one

Exhibit 8.11   Sample Object Definition Table

Obj. No.	Name	Description	Type	Level
1	OBJ 01	Customer	H	Factory
2	OBJ 02	Customer care Web site	D	Factory
3	OBJ 03	Order configuration program	D	Factory
4	OBJ 04	Product database	D	Line
5	OBJ 05	Order management program	D	Line
6	OBJ 06	Order status database	D	Line
7	OBJ 07	Customer database	D	Factory
8	OBJ 08	Finished goods inventory database	D	Line
9	OBJ 09	Purchasing manager	H	Line
10	OBJ 10	Finished goods inventory manager	H	Line
11	OBJ 11	Production manager	H	Line
12	OBJ 12	Materials planning manager	H	Line
13	OBJ 13	Production process database	D	Line
14	OBJ 14	Machining cell controller	M	Cell
15	OBJ 15	Coating cell controller	M	Cell
16	OBJ 16	Assembly cell controller	M	Cell
17	OBJ 17	Supplier	H	Line
18	OBJ 18	Supplier database	D	Factory
19	OBJ 19	Accounts payable manager	H	Factory
20	OBJ 20	Accounts receivable manager	H	Factory
21	OBJ 21	Accounting database	D	Factory
22	OBJ 22	Shipping room team	H	Factory
23	OBJ 23	Assembly cell equipment	M	Machine
24	OBJ 24	Coating cell equipment	M	Machine
25	OBJ 25	Machining cell equipment	M	Machine

vendor, there will be a need to evaluate which components from its product portfolio should be included. It is then necessary to build a database of the capabilities for each solution component that is relevant to the target e-factory scenarios and requirements. Obviously, this database requires two important types of information: (1) information about the solution component, and (2) the key elements of the scenarios. The data about each solution component should be provided by the supplier of the product or by someone who is an expert in the product. The key elements of the scenarios are available from the Scenario Generation Table. A generic table for capturing the necessary solution set component data is illustrated in Exhibit 8.12.

In Exhibit 8-12, Column 1 is the Scenario ID. Column 2 is the activity number for the scenario. Column 3 is the generic type of component being described. Column 4 is the identification number of the component. Column 5 is the control level at which the package operates. Column 6 is the requirement to be satisfied by the solution component. Column 7 is the potential role for the solution component. The role is one of the objects defined in relevant scenario. Column 8 is a score of how well the component

Exhibit 8.12   Solution Component Capabilities Table

Col.1 Scenario ID	Col.2 Activity No.	Col.3 Generic Type	Col.4 Component ID No.	Col.5 Control Level	Col.6 Requirements To Be Satisfied	Col.7 Potential Role	Col.8 Req. Score
1b	3	CRM	3	Factory	Customer care Web pages calls for product configurator program	OBJ02	0.7
1b	4	CRM	3	Factory	Order configurator program takes question and queries product database	OBJ03	0.7
1b	6	CRM	3	Factory	Order configurator program sends answer to customer care Web page	OBJ03	0.6
1b	7	CRM	3	Factory	Customer care Web page sends answer to customer	OBJ03	0.5
1b	9	CRM	3	Factory	Customer care Web page calls for order configurator program	OBJ03	0.5
1b	10	CRM	3	Factory	Order configurator program converts requested configuration into a quotation	OBJ03	0.9
1b	11	CRM	3	Factory	Order configurator program sends answer to customer care Web page	OBJ03	0.9
1b	12	CRM	3	Factory	Customer care Web page sends answer to customer	OBJ03	0.8
1b	14	CRM	3	Factory	Customer care Web page accepts customer order confirmation and sends it to order management program	OBJ03	0.7

satisfies the requirement. The range for the score is from 0 (no satisfaction of the requirement) to 1.0 (complete satisfaction of the requirement). This score can be marked by the suppliers' self-assessment or by the roadmap team after interviewing current users of the component package.

In addition to the requirements defined during the scenario generation process, there will be requirements for interfacing with other solution components. The interface requirements will be defined by the interactions (source, perform, receive) between the objects that operate in the scenario. These interactions can be easily determined from the scenario data by creating an array of interactions. An example of such an array is shown in Exhibit 8.13.

Solution Set Mapping Table

The Soltion Set Mapping Table is used to map the solution set candidates into the scenarios. This is one of the more crucial tables because it captures the thinking about how the solution set technologies can be used and applied at the detailed scenario level. The process for creating the Solution Set Mapping Table is based on consolidating the solution components into the best combinations. A flow chart of this process is shown in Exhibit 8.14. This process establishes the basis for the final evaluations and will ultimately produce the final roadmap and implementation plan.

The Solution Set Candidate Definition Table is a critical piece of the methodology because this is where the consolidation of solution components into solution set candidates is performed. Included in this table is the mapping of the candidate solution set into the object-oriented structure of the entire e-factory. This makes it possible to evaluate all the different combinations of technology providers and the individual packages from those providers.

The key elements in defining an e-factory solution set include:

1. The name of each company with one or more candidate packages
2. The name of each package from each company that is a candidate package
3. The role of each package from each company (role here means the object that it becomes in the scenarios being tested)
4. The control level of each package (this is related to the role; when the role is assigned, the control level is defined as well)

Just the assignment of the above definitions requires some deliberation because it implies the creation of an architecture. Therefore, some type of iteration might be required before a final assignment of roles and definitions is made for each candidate set.

A generic sample solution set table is illustrated in Exhibit 8.15. Column 1 in the table is the identification indicator of the solution set. Column 2 is the identification indicator of each component in the solution set. Column 3 is the name of the generic type of the package. Column 4 is the name of the technology supplier. Column 5 is the name of the component package.

Exhibit 8.13 Scenario Object Interaction Array

Interactions Source	Receive →	1	CRM 2	CRM 3	PDM 4	SCP 5	MES 6	9	SWM 10	11	12	MES 13	FAS 14	FAS 15	FAS 16	17	EPS 18	19	20	ERP 21	22	FAS 23	FAS 24	FAS 25	Total
OBJ	1		4																1						5
CRM	2	2		2			1																		5
CRM	3		2	1	1																				4
PDM	4			1																					1
SCP	5	2																							2
MES	6					1	1			1															3
	9															1									1
WMS	10							1																	1
	11						3	1		1															5
	12								1		1	1													3
MES	13						1				1	1	1	1	1										6
FAS	14									1															1
FAS	15									1															1
FAS	16									1															1
	17												1	1	1										3
EPS	18										1														1
	19															1		1							2
ERP	21						1																		1
FAS	22																				5	1			6
FAS	23																				1	3	1		5
FAS	24																	1					3	1	5
FAS	25															1	1			1				2	5
Total		4	6	4	1	1	7	2	1	5	3	2	2	2	2	3	1	2	1	1	6	4	4	3	67

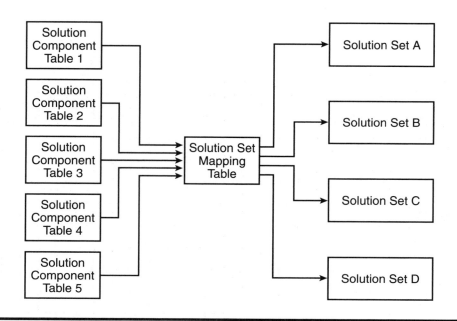

Exhibit 8.14   Creating Solution Sets

Exhibit 8.15   A Sample Candidate Solution Set

Col.1 *Solution* *Set ID*	*Col.2* *Component* *ID*	*Col.3* *Category*	*Col.4* *Company Name*	*Col.5* *Package Name*
A	1	CCM	Cincom Systems, Inc.	Encompass
A	2	CMMS	Datastream Systems, Inc.	MP5/MPXconnect
A	3	CRM	Siebel	Siebel eBusiness 2000
A	4	EPS	Ariba	Ariba Buyer
A	5	ERP	J.D. Edwards	OneWorld
A	6	FAS	Brooks Automation (AutoSimulations, Inc.)	AutoMod
A	7	LMS	J.D. Edwards	OneWorld
A	8	MES	Brooks Automation (FASTech)	FACTORYworks
A	9	PDM	MatrixOne	eMatrix
A	10	SCE	J.D. Edwards	OneWorld
A	11	SCI	BroadVision	MarketMaker
A	12	SCP	i2 Technologies	RHYTHM
A	13	SFA	Siebel	Siebel Sales
A	14	WMS	Catalyst International, Inc.	WMS Release 3

Aditional solution set candidates might include other combinations of these same packages or other packages. Candidate Solution Set C might consist of a configuration similar to Solution Set A (Exhibit 8-15), but with a different enterprise system. A table with other candidate solution sets for the example is given in Exhibit 8.16.

Requirements Value Definition Table

The Requirements Value Definition Table includes the list and definition of the requirements value criteria that are used in scoring how well a solution set satisfies a set of requirements. While such a table may not be terribly exciting in terms of its content, it does act as documentation of the degree of granularity that is being used in the solution set evaluation. The example table given in Exhibit 8.17 shows an 11-level criteria table. Any scheme using fewer or more levels can be applied. Of course, with more levels in an evaluation scheme, there can be greater differentiation in an evaluation but greater effort may be required to assess each requirement. For example, as each solution set candidate is evaluated as to how well it satisfies a requirement, a number from 0.0 to 10 is chosen from the Requirements Definition Table to reflect the level of requirement satisfaction.

Cost Value Definition Table

The Cost Value Definition Table includes the list and definition of the cost value criteria that are used in evaluating the value of a solution set when compared to other solution sets. Similar in purpose to the Requirements Value Definition Table, it also acts as documentation of the degree of granularity that is being used in the solution set evaluation.

However, the definition of value used in this table is different from that in the other value definition tables. The definition of value used here is the same as that used in value engineering techniques. For example, if there are three solution set components which have the same functionality but which have different costs (C1, C2, and C3), then the lowest cost solution set (say, C2) has the highest value. The other components have higher costs and, therefore, less value. If each cost is divided by C2, value ratios between 0.0 and 1.0 are created. Maximum value is 1.0, which means that no other alternative is less costly. The smaller the value of a sample, the more costly it is in comparison to the lowest cost alternative.

If the cost value is calculated using the ratios of actual costs, then no table is required. If the value is difficult to calculate because there is no specific cost data available, a more subjective approach may be required by making qualitative assessments about the relative costs. The sample table shown in Exhibit 8.18 shows an 11-level value table. Any scheme using fewer or greater numbers of levels can be applied. Of course, with more levels in an evaluation scheme, there can be greater differentiation in an evaluation but greater effort may be required to assess or to conclude on the assessment for each cost comparison.

Exhibit 8.16 Candidate Solution Set List

Category	Solution Set A	Solution Set B	Solution Set C	Solution Set D
CCM	Cincom Systems, Inc.	Voxco	Cincom Systems, Inc.	Cincom Systems, Inc.
CMMS	Datastream Systems	Indus Int., Inc.	Fluor Daniel, Inc.	Fluor Daniel, Inc.
CRM	MarketForce	Siebel	Siebel	MarketForce
EPS	Ariba	Trilogy Software	Commerce One	Ariba
ERP	J.D. Edwards	SAP	Oracle	J.D. Edwards
FAS	AutoSimulations, Inc.	Camstar Systems, Inc.	Camstar Systems, Inc.	AutoSimulations, Inc.
LMS	J.D. Edwards	TRW	Manugistics, Inc.	TRW
MES	Brooks Automation (FASTech)	Invensys	Emerson Electric (Intellution)	Brooks Automation (FASTech)
PDM	Autodesk Inc.	CONCENTRA	Autodesk, Inc.	Autodesk Inc.
SCE	J.D. Edwards	Manhattan Assoc., Inc.	Manhattan Assoc., Inc.	Manhattan Assoc., Inc.
SCI	BroadVision	Extricity	BroadVision	BroadVision
SCP	I2 Technologies	Manugistics, Inc.	J.D. Edwards	i2 Technologies
SFA	Infinium	Infinium	Infinium	Infinium
WMS	Catalyst Int., Inc.	TRW	Manhattan Assoc., Inc.	TRW

Exhibit 8.17   Requirements Value Definition Table

Criteria	Value	Degree Satisfied
0	—	0%
1	0.10	0–10%
2	0.20	10–20%
3	0.30	20–30%
4	0.40	30–40%
5	0.50	40–50%
6	0.60	50–60%
7	0.70	60–70%
8	0.80	70–80%
9	0.90	80–90%
10	1.00	100%

Exhibit 8.18   Sample Cost Value Definition Table

Criteria	Value	Cost Value Calculation Range
0	—	0%
1	0.10	0–10%
2	0.20	10–20%
3	0.30	20–30%
4	0.40	30–40%
5	0.50	40–50%
6	0.60	50–60%
7	0.70	60–70%
8	0.80	70–80%
9	0.90	80–90%
10	1.00	90–100%

With fewer levels, there is less granularity in the evaluation, but the cost-by-cost assessment may be easier. This table merely serves as documentation of the scheme used.

Time Value Definition Table

The Time Value Definition Table contains the list and definition of the criteria for evaluating the value of the time effectiveness of a solution set. The value formula for calculating a time is similar in structure to that for calculating a cost value; that is for each solution set candidate:

Candidate time value = Lowest solution set cycle time / Candidate cycle time

The example table in Exhibit 8.19 shows an 11-level value table. Any scheme using fewer or greater numbers of levels can be applied. Of course,

**Exhibit 8.19   Time Value
Definition Table**

Criteria	Value	Time Value Ratio
0	—	0%
1	0.10	0–10%
2	0.20	10–20%
3	0.30	20–30%
4	0.40	30–40%
5	0.50	40–50%
6	0.60	50–60%
7	0.70	60–70%
8	0.80	70–80%
9	0.90	80–90%
10	1.00	100%

with greater levels in an evaluation scheme, there can be greater differentiation in an evaluation but greater effort may be required to assess or to conclude on the time value of each solution set. With fewer levels, there may be fewer shades of gray distinguishable in the evaluation, but the time value assessment may be easier. This table merely serves as documentation of the scheme used.

Solution Set Evaluation Table

A composite total value of a candidate solution set can be created from the product of the requirements value, cost value, and time value as defined in the following equation:

Total value = Requirements value * Cost value * Time value

This combination of values allows for a compound evaluation in terms of the three values of requirements satisfaction, low cost, and rapid action. However, this total value calculation can be modified to emphasize a reduced set of criteria or a broader set. If cost is of no concern, then the cost factor can be eliminated (the equivalent of setting cost value to 1.0 for all solution set candidates) and the total value calculation would consist of just two factors (i.e., requirements value and time value). If time is not an issue, then set the time value parameter to 1.0.

The Solution Set Evaluation Table is the summary of all value calculations of how effectively each solution set candidate satisfies each requirement, cost, and cycle time goal. An example of this table is given in Exhibit 8.20. This table contains the final scores for all the candidates and thus becomes the

Exhibit 8.20 Solution Set Evaluation Table Example

SolutionSet	CostValue	TimeValue	Req.Value	TotalValue
A	0.83	0.74	0.89	0.55
B	0.91	0.54	0.88	0.43
C	0.35	0.64	0.93	0.21
D	0.67	0.32	0.63	0.14

database from which the best solution set can be identified. In the example in Exhibit 8.20, the best solution set is Solution Set A because it has the highest total value.

Creating this table forces the team making the decision to clearly go on record as to what it views as the most important factors in making a decision. If the team decides that cost should not be a final decision factor but that process speed and functional performance should be, then this can be implemented by making the cost value 1.0 for all solution sets. On the other hand, if an evaluation goal is that all the solution sets above a certain level of performance are viable so that the final decision is made only on price, then the numbers in the requirement and time value tables for the solution set candidates that pass the performance levels should be adjusted to 1.0.

Implementation Requirements Table

The Implementation Requirements Table contains the list and definition of the fundamental steps necessary to implement the e-factory solution set selected to be the best set. This information is important in helping determine how much time and effort will be required to bring different portions of the final total solution set into operation. This helps define the final roadmap. An example of this table is given in Exhibit 8.21.

In Exhibit 8.21, Column 1 is the number of the implementation step. Column 2 is the description of the implementation step. Column 3 is the number of workdays required to perform the implementation step. Column 4 is the identification of the object whose role the solution set component will fulfill. Column 5 is the identification of the solution set. Column 6 is the identification of the component from the solution set that is being implemented. Column 7 is the number of full-time equivalent people (FTEs) required to execute the implementation step. Column 8 is the total number of person-days required to execute the implementation step. This is the result of the number of FTEs multiplied by the number of workdays for the implementation step. Column 9 is the cost of the person-days, which is the result of the multiplication of Column 8 by the cost per person-day. Column 10 is the phase of the implementation program in which the implementation step falls because of precedence, priority, or policy.

Exhibit 8.21 Example of an Implementation Requirements Table

Col.1 Imp. Step	Col.2 Description	Col.3 Imp.Time (days)	Col.4 Object	Col.5 Solution Set	Col.6 Component	Col.7 FTEs	Col.8 FTE Days	Col.9 People Cost	Col.10 Phase
1	Design business process	20	OBJ 01	A	1	2	40	$16.0	1
2	Select and acquire materials	5	OBJ 01	A	1	1	5	$2.0	1
3	Design software configuration	20	OBJ 01	A	1	1	20	$8.0	1
4	Build software configuration	30	OBJ 01	A	1	2	60	$0.0	1
5	Cleanse and implement data	40	OBJ 01	A	1	2	80	$32.0	1
6	Train staff	20	OBJ 01	A	1	1	20	$8.0	1
7	Test implementation	20	OBJ 01	A	1	2	40	$16.0	1
8	Go live	5	OBJ 01	A	1	1	5	$2.0	1
9	Get feedback	40	OBJ 01	A	1	1	40	$16.0	1
10	Improve implementation	20	OBJ 01	A	1	2	40	$16.0	1

The implementation requirements factor into determining the most practical schedule and project plan for introducing the selected solution set components into the business processes, as well as introducing new process activities. This table should include data for each object and each solution set component being evaluated. When completed, this table can then be used to create the Solution Set Roadmap Table.

Solution Set Roadmap Table

The Solution Set Roadmap Table summarizes the results of all the previous steps in the process. It is created by taking the data in the Implementation Requirements and organizing it into phases and implementation types. There are at least three ways to present the results of the Solution Set Roadmap Table. One form of the table, that which shows the effort required to implement the e-factory in labor-hours, is shown in Exhibit 8.22. Another form of the table that shows the cost required to implement the e-factory in dollars is shown in Exhibit 8.23. A third form of the table, which shows a Gant chart of the time scale of the implementation, is shown in Exhibit 8.24.

The focus of the Solution Set Roadmap Table is on implementation of scenario objects. As previously discussed, the entire methodology is based on the definition of scenarios and the objects that operate within each scenario. An object can be human, machine, or data. If the object is machine or data, then implementing the roadmap means implementing a technology object. If the object is human, then implementing the roadmap means implementing training programs that teach the people — the human object — how to operate in the scenarios.

In Exhibit 8.22, Column 1 denotes the type of implementation (technology or training) for each object. Column 2 is the name of the object being made possible by the implementation. Column 3 is the object type (human, data, or machine). Column 4 is the identity of the solution set. Column 5 is the identity of the component of the solution set being implemented as part of the object. Column 6 is the effort that would be expended in phase 1. Columns 7, 8, and 9 include the efforts that would be expended in phases 2, 3, and 4.

In Exhibit 8.23, Columns 1 through 5 represent the same information as in Exhibit 8.22. Columns 6 through 9 represent the cost of implementing technology or training for each object by phase; and Column 10 is the sum of all the costs by object.

In Exhibit 8.24, the implementation roadmap by project and by time scale is shown in a Gant chart format. There is no new data in this format, but it is a useful way for the roadmap team or steering committee to discuss the key elements of the e-factory roadmap.

Completing the roadmap for this example results in several conclusions in addition to the roadmap. For example, the roadmap methodology for this

Exhibit 8.22   Solution Set Roadmap Table (Effort)

Col.1	Col.2	Col.3	Col.4	Col.5	Col.6	Col.7	Col.8	Col.9
Imp. Type	Object Name	Object Type	Solution Set	Solution Component	Phase 1	Phase 2	Phase 3	Phase 4
Technology	OBJ 02	D	B	3				80
	OBJ 03	D	B	3				80
	OBJ 04	D	B	9	60			
	OBJ 05	D	B	10		100		
	OBJ 06	D	B	8	60			
	OBJ 07	D	B	3				60
	OBJ 08	D	B	14		80		
	OBJ 13	D	B	8	80			
	OBJ 14	M	B	6	80			
	OBJ 15	M	B	6	80			
	OBJ 16	M	B	6	60			
	OBJ 18	D	B	4			100	
	OBJ 21	D	B	5			60	
	OBJ 23	M	B	6	100			
	OBJ 24	M	B	6	60			
	OBJ 25	M	B	6	80			
	OBJ 26	M	B	8	80			
	OBJ 27	D	B	7		100		
	OBJ 28	D	B	2	60			
	OBJ 30	M	B	6	80			
	OBJ 31	D	B	6	60			

Training						
OBJ 09	H	B	3			20
OBJ 10	H	B	3			15
OBJ 11	H	B	3			15
OBJ 12	H	B	4		20	
OBJ 17	H	B	5		20	
OBJ 19	H	B	6	15		
OBJ 20	H	B	4		15	
OBJ 22	H	B	5		10	
OBJ 29	H	B	6	20		

Exhibit 8.23　Solution Set Roadmap Table (Cost)

Col.1	Col.2	Col.3	Col.4	Col.5	Col.6	Col.7	Col.8	Col.9	Col.10
Imp Type	Object Name	Object Type	Solution Set	Solution Component	Phase 1	Phase 2	Phase 3	Phase 4	Grand Total
Technology	OBJ 02	D	B	3				$96	$96
	OBJ 03	D	B	3				$64	$64
	OBJ 04	D	B	9	$96				$96
	OBJ 05	D	B	10		$120			$120
	OBJ 06	D	B	8	$96				$96
	OBJ 07	D	B	3				$96	$96
	OBJ 08	D	B	14		$64			$64
	OBJ 13	D	B	8	$96				$96
	OBJ 14	M	B	6	$128				$128
	OBJ 15	M	B	6	$96				$96
	OBJ 16	M	B	6	$48				$48
	OBJ 18	D	B	4			$120		$120
	OBJ 21	D	B	5			$48		$48
	OBJ 23	M	B	6	$160				$160
	OBJ 24	M	B	6	$72				$72
	OBJ 25	M	B	6	$64				$64
	OBJ 26	M	B	8	$64				$64
	OBJ 27	D	B	7		$80			$80
	OBJ 28	D	B	2	$96				$96
	OBJ 30	M	B	6	$128				$128
	OBJ 31	D	B	6	$72				$72

Training								
OBJ 09	H	B	3				$24	$24
OBJ 10	H	B	3				$18	$18
OBJ 11	H	B	3				$18	$18
OBJ 12	H	B	4			$24		$24
OBJ 17	H	B	5			$16		$16
OBJ 19	H	B	6		$12			$12
OBJ 20	H	B	4			$30		$30
OBJ 22	H	B	5			$20		$20
OBJ 29	H	B	6	$24				$24
Grand Total				**$1252**	**$264**	**$258**	**$316**	**$2090**

ID	❶	Task Name	M1	M2	M3	M4	M5	M6	M7	M8	M9	M10	M11	M12	M13	M14	M15	M16	M17
				4th Quarter			1st Quarter			2nd Quarter			3rd Quarter			4th Quarter			1st
1		**Phase 1**																	
2		OBJ 19																	
3		OBJ 29																	
4		OBJ 04																	
5		OBJ 06																	
6		OBJ 16																	
7		OBJ 24																	
8		OBJ 28																	
9		OBJ 31																	
10		OBJ 13																	
11		OBJ 14																	
12		OBJ 15																	
13		OBJ 25																	
14		OBJ 26																	
15		OBJ 30																	
16		OBJ 23																	
17		**Phase 2**																	
21		**Phase 3**																	
28		**Phase 4**																	

Exhibit 8.24   Solution Set Roadmap Table (Schedule)

example concludes that the generic solution component candidates SFA, SCE, SCI, and CCM are not needed. Another conclusion is that objects OBJ 01, OBJ 09, OBJ 10, OBJ 11, OBJ 12, OBJ 17, OBJ 19, OBJ 20, OBJ 22, and OBJ 29 are not technology objects.

The Solution Set Roadmap Table is the final step in building a comprehensive database of the assumptions, business process design decisions, supply chain requirements, and technology candidates. The tables illustrated in this chapter are samples of the data contained in the example database.

Summary

After a roadmap team has been formed and a vision of the e-factory created, a roadmap for the e-factory can be built. The key initial step in the roadmapping methodology is the collection of the requirements from the supply chain of customers, suppliers, and operations internal to the e-factory. An equally important initial step is the creation of the scenarios that describe the full range of operations to be experienced and performed by the e-factory. Each scenario is composed of the objects and activities that define the interactions of the objects and the flow of material and information through the business process. After the scenarios are created or in parallel with their creation, the candidates for the technology solution sets need to be selected. A solution set can be composed of more than one solution component. One solution component can be part of more than one solution set. The candidate solution sets then need to be mapped into the scenarios so that the evaluation of the solution sets can be completed. Mapping the solution sets into the scenarios means that each of the non-human objects needs to be assigned to a solution set component, and vice versa.

Once the solution set components are mapped into the scenarios, then each solution set can be evaluated on the basis of how well it satisfies requirements, minimizes cycle time, and/or minimizes cost. This evaluation should result in a "best" solution set based on these criteria. Once the best solution set is identified, then its implementation requirements can be used to create a roadmap which has the resources, time, and cost organized into a phased time schedule. If it is desired to include implementation time or cost as an evaluation criterion, the evaluation step can be performed with implementation values. A new, expanded solution set evaluation table can then be created for each of the final candidates so that a final decision can be made on the basis of implementation time and cost. Otherwise, if a best solution set is defined by the evaluation tables, then only one Solution Set Roadmap Table needs to be created.

The primary purpose of the roadmap methodology is to ensure that an e-factory can be defined and its implementation planned in the new and rapidly evolving environment. This methodology is designed to deal with many factors, many requirements, and many alternative solutions. It provides for the rapid structuring of issues and forces the evaluation of key factors. In this way, the roadmap methodology should maximize the quality of the decisions being made.

Chapter 9

Creating the e-Factory Roadmap

Purpose of This Chapter

Chapter 8 focused on methodology. However, a methodology is useful only when there are people working together and using the methodology. The purpose of this chapter is to describe an approach for getting all the appropriate people involved in the process of creating the e-factory roadmap.

Outline of the Process

The best approach to creating the e-factory roadmap is also the simplest and best-tested approach. The approach is based on establishing a team of people from all the organizational functions involved in the key e-factory business processes. This team of people then works together using the methodology described herein to ensure that all the necessary information is gathered and that all the relevant decisions are made. The approach consists of five simple steps, as illustrated in Exhibit 9.1.

This team approach is very similar to the approaches used by other methodologies such as Just-in-Time (JIT), Total Quality Management (TQM), Business Process Reengineering (BPR), Six Sigma (6 sigma), Change Management, Quality Function Deployment (QFD), and Value Engineering (VE). The reason for a team of professionals is that the combined intelligence of the team and the work output of a coordinated team will yield more than the accumulated intelligence and effort of the same number of individuals. Each of these steps and how the team operates in each are described in the following.

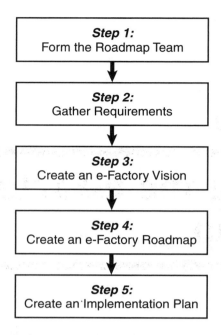

Exhibit 9.1   e-Factory Roadmap Creation Process

Step 1: Forming the Roadmap Team

As stated, the e-factory roadmap should be built by a cross-functional team. Although it is usually better to have a team of people working together in general, the cross-functional team is absolutely essential in this case because of the need to build business process scenarios. The scenarios are critical to a successful e-factory roadmap, and scenarios can only be created effectively by the people who will be involved in the design and operation of the business process. The key steps in forming the roadmap team are described in the following paragraphs.

Team Selection

Selection of e-factory team members should be done by a management team that should include the most senior executive who should be the company champion, the human resource manager, and other executives who are involved in the e-factory processes. Team members are selected from existing functional organizations. The size and composition of the team should reflect the minimum skill set required to successfully execute the e-factory methodology. However, there are limits to team size. The team should ideally consist of a core group and a support group. The core group should be dedicated, co-located, and consist of members from the relevant functions.

The team should be selected on the basis of the type of business process being supported by the roadmap. For an order fulfillment process, the cross-functional team should consist of representatives from:

- Process engineering
- Production control
- Materials management
- Selected suppliers
- Selected customers
- Information technology
- Supply chain management
- Purchasing
- Customer service
- Field sales
- Marketing

A product realization process would consist of representatives from many of the same functions and include:

- Selected suppliers
- Selected customers
- Information technology
- Supply chain management
- Purchasing
- Customer service
- Field service
- Product marketing
- Engineering (e.g., systems, applications, test, process)
- Research and development (R & D)
- Manufacturing (e.g., industrial engineering, process planning, etc.)

The support group for the core team does not necessarily have to be dedicated full-time to the team although when their services are needed, they should be available for as long as needed. The support group typically includes members whose skills are not needed except for special questions or activities.

Formal steps should be taken to ensure that all appropriate organizational functions be involved throughout the life of the project, rather than being informed at the end of their accomplishments. This implies that all members of the team will be involved in decision-making from the beginning to prevent decisions from being reworked if the job were passed from one functional organization to another as with traditional approaches.

Functional managers must recognize the priority of tasks assigned to team members and must subordinate other activities that require the team member's attention or involvement. Also, team members should receive a performance review that reflects their performance as a team member.

The team should have clear ownership of the e-factory roadmap project and process concept decisions. Team members must avoid any functional organization loyalties that negatively affect the team decision-making process. By bringing together all the people who can affect the innovation of an e-factory and empowering them to make decisions, teams will be able to rapidly develop the e-factory roadmap.

Team Leadership

Given an empowered, cross-functional team, the team leader has a special role during the life of a project. A team leader is assigned when the team is formed, and is one of the working members of the team. The role of team leader should be one of support and team building, rather than control. The team leader should embody the following characteristics.

1. Have an understanding of the overall business aspects of the product line or industry segment being served
2. Have an understanding of the technologies being considered for use by the team
3. Be able to work with people from different cultures, language backgrounds, technical disciplines, and organizational functions
4. Be able to manage conflicting ideas or recommendations
5. Be able to facilitate team consensus
   a. On agreement on key issues
   b. On disagreements requiring management decision
6. Be able to articulate issues and positions clearly and fairly
7. Be able to sustain a sense of commitment and urgency with the team

The responsibilities and duties of the team leader include:

1. Project management, including the development and tracking of all schedules and costs against targets
2. Communicating the team's progress to management
3. Facilitating the team decision-making process and serving as arbiter in resolving conflicts
4. Responsibility for securing the resources the team requires to complete all development activities
5. Ensuring that management challenge and market requirements are constantly being represented in the roadmap creation process
6. Promoting value-adding activities from all team members
7. Representing the team decision process when making recommendations to management on key issues

Team Empowerment

Empowerment of the e-factory roadmap must come from company management and the executive in charge of the team. However, the team leader will have responsibility for the day-to-day conduct of team affairs and will have signatory authority and responsibility. Ideally, the team should be empowered to make all decisions necessary to complete its mission in the best interest of the company and its customers.

Agreeing to Team Goals and Metrics

The team is accountable for the creation of an e-factory roadmap. Specific performance metrics are used to monitor progress and to measure final results. The metrics used are determined and agreed upon by all team members and must include measures for thoroughness, quality, timing, and efficiency. Metrics should encourage teamwork for creating an e-factory roadmap for the business rather than optimizing functional preferences or old paradigms that may have been the norm previously. Useful metrics include:

- On-schedule completion
- Cost to create the roadmap
- Satisfaction of customer, supplier, and internal process requirements

In addition to establishing team goals and metrics, individual team members should also establish their personal goals within the project and the metrics to monitor achievement of these goals. The team should create metrics to gauge performance over the length of the project. This will include near-term metrics that are related to each project phase, as well as those aimed at the overall innovation project.

Performance rewards for team members should be primarily related to team performance against business-oriented metrics. These should be agreed upon by management and the team. Individuals should not be rewarded on the basis of their functional performance, but rather on the basis of their participation in an innovation project that is successful from a business standpoint.

Methodology Sign-off and Review

Prior to the methodology sign-off, the team will have received and reviewed the management challenge to create an e-factory roadmap and translated the challenge into specific e-factory goals. At sign-off, the team accepts ownership of its challenge, which are business goals translated into team goals that are measurable. Participants at the sign-off should use a checklist such as the one shown in Exhibit 9.2 to ensure completeness when establishing consensus on the requirements and any necessary changes.

Signature of the team members represents their approval of the team challenge and their commitment to work as a team. Establishing a cross-functional team is the single most important decision to be made for an e-factory roadmap creation project and requires a clear set of guidelines and operating principles.

Exhibit 9.2   Methodology Sign-off Checklist

Team Name:_____   Date:_____

    1. Does the team accept the project cost target of $_____?
       YES_____   NO_____
    2. Does the team accept the roadmap goals?
       YES_____   NO_____
    3. Does the team accept the project schedule with completion date_____ ?
       YES_____   NO_____
    4. Are all critical goals satisfactorily described?
       YES_____   NO_____   If not, list which ones:
    5. Do any of the goals conflict with each other?
       YES_____   NO_____   If so, list below:
    6. Do the goals conform to the original management challenge?
       YES_____   NO_____   If not, explain what actions will be taken.
    7. Have individual and team appraisal criteria been developed?
       YES_____   NO_____   If not, explain what actions will be taken.
    8. Does each team member understand how he/she will be measured?
       YES_____   NO_____   If not, explain what actions will be taken.

Step 2: Gathering the Requirements

As described in previous chapters, gathering the e-factory requirements is the most important part of the total process. The requirements must come from three areas: those driven by customer needs, by supplier needs, and by the process needs internal to the company. Gathering requirements can be done in several ways, through surveys, interviews, brainstorming sessions, and focus groups (see Exhibit 9.3).

Brainstorming sessions are important for cross-functional teams to free their thinking and to get everyone to contribute to the thinking and analysis process. The purpose of the brainstorming session should be to identify as many ideas about what the requirements "could" be. The term "brainstorming" itself highlights the fact that these sessions are designed to get people to use their

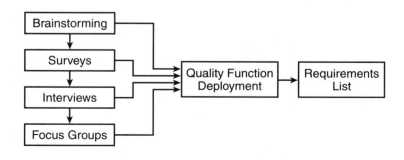

Exhibit 9.3   Gathering Requirements

imagination of what could be possible and memory of what they have seen in other situations to suggest possible requirements. These brainstormed requirements should be viewed only as a starting point of requirements to be tested along with other datagathering activities such as interviews and focus groups. If the brainstorming team also includes customer and supplier representatives, the brainstorming activity also helps reveal some of the issues that they face everyday. These issues could be opportunities for the new e-factory roadmap to pursue. The output of the brainstorming sessions can be used as material to construct surveys and interviews as well as final requirements.

Surveys are useful for gathering requirements from customers and suppliers on a broad scale. If the markets being served are mass markets in nature with a large number of customers, then electronic, telephonic, or written surveys can be used to try to cover a statistically significant number of customer requirements. For industries that serve a small number of customers or several large customers, face-to-face interviews are more effective. The output of surveys can be used as material to construct or refine interview guides for the face-to-face interviews as well as the final requirements.

Face-to-face interviews are useful for gathering requirements from key customers or key individuals inside of the key customers. Interviews should be structured so that the interview time can be spent efficiently. However, the interviewer should be trained to allow the course of the interview to flow into areas that the customer considers important even if the pre-interview preparation may not have identified them. The interview process should be interactive. The interview structure should be reviewed and revised after each set of interviews to ensure that any new interviewing techniques or issues can be quickly incorporated for future interviews. The output of the interviews can be used as material to construct focus group as well as final requirements.

Focus groups can be useful when there are very specific issues that have to be determined in an uncertain environment. If there is confusion or conflict in the requirements gathered by other means, then a focus group may be needed. A focus group of people from one or more of the three key requirements source groups is also a useful way to get a final determination of the true evaluation of the interaction and prioritization of the gathered requirements. Focus groups require significant preparation and skillful facilitation to ensure that group members respond from actual feelings and are not biased by the format or content of the questions asked.

After all the data is collected, it must be transformed into a list of requirements that reflect the specific needs identified during the data gathering. The list should be prioritized, and should have specific target values if possible. One of the best methods for this is a technique called quality function deployment (QFD). Developed more than 20 years ago by people in Japan and the United States, the technique was originally intended to be an organized way to design products and processes by forcing a cross-functional team to compare and contrast quality or functional requirements and specify solutions to those requirements. Over time, it has been used to facilitate requirements analysis and decision-making.

The QFD tool was invented in Japan during the 1970s. It resulted from the desire of government-sponsored researchers in Japan to capture in analytical form the best practices of the successful Japanese manufacturing companies. Although developed there, QFD was actually not practiced extensively in Japan. Also it is probably misnamed because of its generic capability. However, it was brought to the United States by the Ford Motor Company who found it to be a useful tool for product development and process design in bringing together the many "voices" in its operations to form one voice — the voice of the customer. Since its inception and introduction in the United States, its use has evolved to the point that it can become a major factor in reducing the cycle time of making decisions about requirements for new products and new processes. It assists in ensuring the capture of the "voice of the customer" as well as the "voice of the supplier" and the "voice of internal operations."

QFD can be an important factor in defining a set of requirements that originate from a variety of sources. It provides a structure for organizing the thoughts, impressions, and data that may exist in different functions in an organization. It provides a means by which a multi-functional team can work together early in a product development program to form a consensus on what the product should be. It provides the basis of a common language or communication mode for the multi-functional team.

The Q stands for quality, which in the simplest of terms is defined to be whatever makes the customer happy; it is the satisfaction of the customer's needs; it is the delivery of what the customer wants. It is easy to talk about what quality is not: it is not products that do not work properly or business processes that are inefficient or ineffective. Quality should be defined in terms of customer expectations met or exceeded.

This definition of quality is, of course, extremely broad. However, by using the broadest definition possible for quality, all of the issues associated with the success of a product or process can be incorporated. For example, a product or process should be designed to provide only the quality for which a customer is willing to pay. A product or process should be designed to provide the quality as it is perceived by the customer, as well as to provide quality surprises when the customer has an unfulfilled desire or expectation.

Product and process are being used interchangeably in this description of QFD because of the value that the QFD approach has in facilitating the collection of requirements. This is important because in the era of e-commerce, the process is, in many ways, the product. With the rapid response required from the e-factory and the enlarged customer experience that the e-factory must provide, the process by which a product is presented and provided to a customer will determine the degree of satisfaction that a customer has just as much as the product itself.

To help explain this concept, consider the simple example of the home kitchen device usually called a bread toaster. Assume that extensive market surveys have been performed and it has been discovered that the typical customer expects a toaster to produce a crunchy brown bread for breakfast. To delve deeper into more detail about this expectation, questions would be asked about safety, cost, exterior shape, messiness, degree of control, number of slices of bread that an be accommodated, types of bread that can be accommodated, and other items that a customer might want to toast. It is all of these requirements that define expected quality. Addressing process issues, questions would be asked about how likely a customer is to buy on-line, what they would expect during that buying experience, and what type of response and service would they like or expect after making a purchase order. For the e-factory, other questions about the speed of response from the retail experience to delivery and usage should be included. With such a broad definition of quality, it is often difficult for two people on the same roadmap team, much less an entire organization, to have the same definition and concept of quality. The QFD enabling tool provides a valuable mechanism for achieving a common concept and common set of goals within a cross-functional team and within an organization.

The F stands for function, that is, what function must be provided by the product or process to satisfy the customer's expectation. Where quality was defined to be the customer's expectations, function is the set of measurable factors or processes that are necessary to satisfy those expectations.

The D stands for deployment — deployment as in how the quality functions are deployed in the product (in the form of parts or features) or in the production process. Deployment is critical because it is the manifestation of decision-making. There can be much discussion and a wide variety of opinions within the team of quality and function, but deployment is where the decisions have to be made as to how the product will be realized to satisfy customer requirements.

The use of QFD at the start of a project is important because it forces the cross-functional issues to be identified and addressed. The tool acts as a forcing function by requiring that a consensus be formed on customer, supplier, and internal requirements; that a consensus be formed on what the technical functions must be to be successful; and that the e-factory strategy of the company be clear and integrated into the roadmap.

The output of QFD is a requirement matrix that summarizes the best thinking and research of a cross-functional e-factory roadmap team. However, it is not so much the matrix itself that is valuable as it is the discussion, the research, and the effort that goes into creating the matrix. QFD represents a disciplined approach to collecting and analyzing data and defining requirements about the e-factory that should be developed to satisfy the market needs. Successful companies go through most of the requirements generating activities called for by the QFD process anyway. The benefit of using QFD is that any company can force itself to perform the analysis and decision-making that the best companies do and do it as early as possible.

The goal of the QFD analysis is to have the voices of the customer, the supplier, and internal operations permeate and promulgate throughout all the activities associated with creating the e-factory roadmap. This is based on the theory that what the customer wants can be defined, and that every activity associated with the business processes that get the product to market should be driven by the requirement of satisfying those wants. An example of a QFD matrix that would be useful to a roadmap team is given in Exhibit 9.4.

Exhibit 9.4   The Requirements QFD Matrix

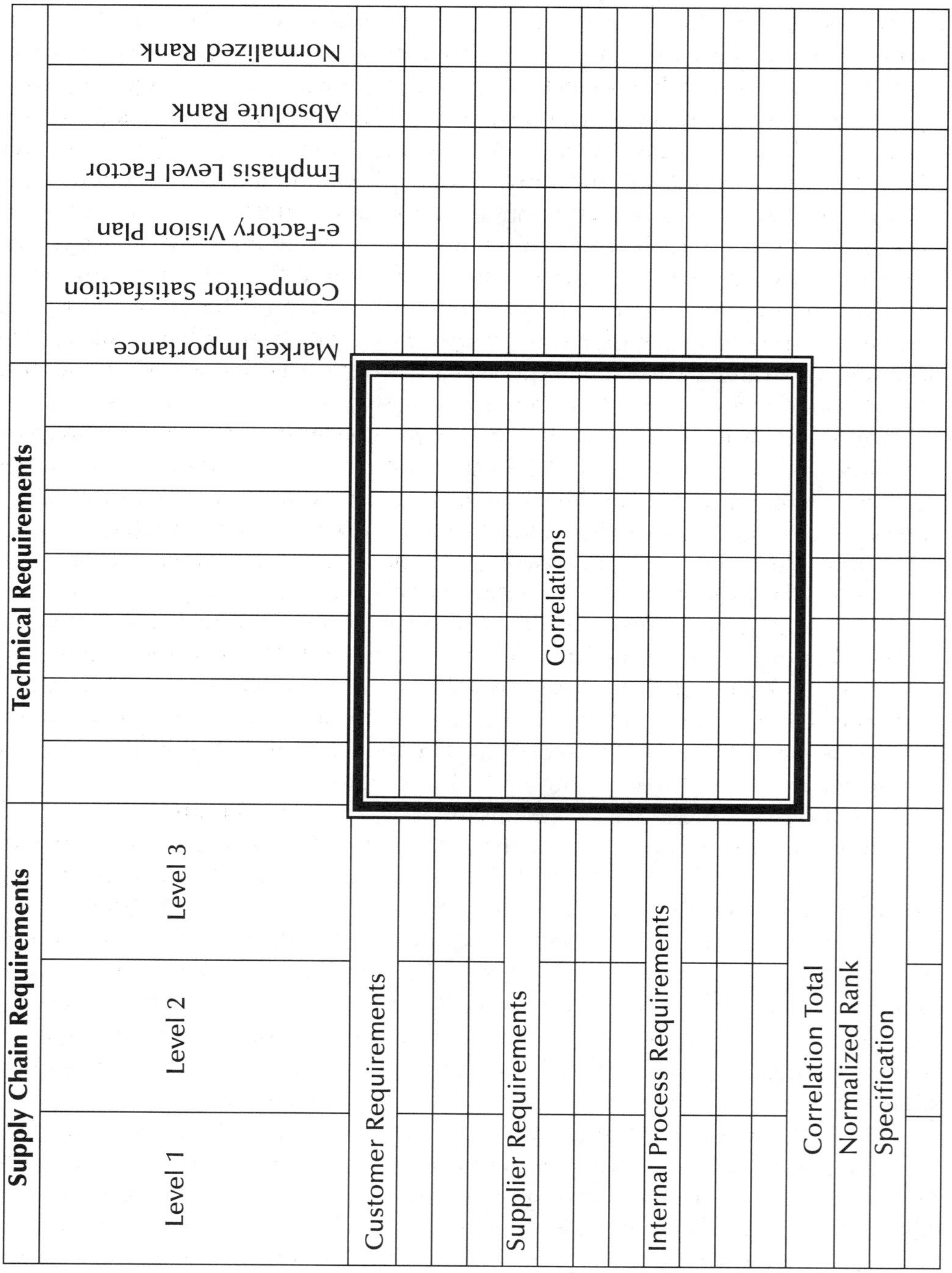

Supply Chain Requirements			Technical Requirements	Market Importance	Competitor Satisfaction	e-Factory Vision Plan	Emphasis Level Factor	Absolute Rank	Normalized Rank
Level 1	Level 2	Level 3							
Customer Requirements			Correlations						
Supplier Requirements									
Internal Process Requirements									
Correlation Total									
Normalized Rank									
Specification									

Constructing a QFD Matrix

The process of building a QFD matrix can be delineated in a straightforward fashion. Performing the data gathering and analysis takes effort. However, the questions asked as part of the QFD activity should be asked as part of an interview, focus group, or brainstorming process. The difference is that the QFD activity is organized in a very disciplined way and forces the questions to be asked early and to be answered by a cross-functional team. The QFD data capture and documentation approach has been electronically integrated into the database tools used for the e-factory roadmap process.

Presented in the following is an example of how to build the first QFD matrix. The steps involved in building the matrix are easy to explain; answering the questions they ask is the challenging part. Exhibit 9.4 is referred to frequently in the following discussion. Each step necessary to construct the first QFD matrix is discussed. By walking through each step, it is possible to see how much decision-making must be done at the earliest stages of a roadmap project by a cross-functional team.

List the Supply Chain Requirements

This is obviously the first and often the most difficult step. This list represents the all-important elements of the voice of the supply chain, that is, the voice of the customer, supplier, or internal business process. This list should contain all the desires and needs perceived by the customer, supplier, or internal process. It can also contain items that have not yet been perceived as being needed, and items that could be viewed as a quality surprise or unexpected benefit. Demands tend to fall into the following types:

1. Physical demands, such as:
   a. Cycle time
   b. Reliability, durability, frequency of breakdown, consistency
   c. Service performance
   d. User friendliness
   e. Safety
   f. Style
   g. Cost
   h. Volume, lot size, shipment size
   i. Packaging
2. Emotional demands, such as:
   a. Esteem
   b. Loyalty
   c. Perceived value
3. Information demands, such as:
   a. Order status
   b. Quality status
   c. Flexibility in content or performance

    d. Access time for information
    e. Data security
    f. Failsafe operation

Although the physical and information demands are the easiest to design for, the emotional requirements can be the deciding factor.

As the requirements data is collected, it should be organized in hierarchical fashion using the format illustrated in Exhibit 9.4. This allows for clearer and more careful consideration by the team during the remaining QFD activities. For example, a first-level customer requirement might be "rapid quotation response." The second level for this first-level requirement might be "shopping cart feature required." The shopping cart feature requirement might have a third-level requirement, which is "prevent products that do not operate with what is already in the shopping cart from being placed in the cart." These demands or supply chain requirements should be placed in the vertical list.

Generate the Technical Requirements

Technical requirements are those technical parameters necessary to satisfy or provide the satisfaction of the customer requirements. Technical requirements must be measurable or definable in measurable terms. One technical requirement may be applicable to many customer requirements. The team generates the technical requirements by considering the customer requirements and articulating the necessary technical requirements. The technical requirements must be measurable because they will be used to define the technical specifications for the e-factory business process or technology.

Using the example above, a technical requirement for the "prevent products that do not operate with what is already in the shopping cart from being placed in the cart" customer demand would be "product configuration database" and "product configuration analyzer." As the technical requirements for each demand are identified, they should be placed in the horizontal list.

List the Market Importance for Supply Chain Requirements

One of the benefits of the QFD matrix approach is that it requires that there be a ranking of all the key factors. For supply chain requirements, a rating or ranking provides a means to discriminate between the more important and lesser important items. The ranking is important because it gives the team a quantified measure of the importance of customer requirements that can be useful if trade-offs have to be made about process or technology features or cost.

The ratings for each customer requirement can be generated in several different ways, including:

- Included as part of original data gathering surveys
- Included as part of focus group activity
- Assessed as a team exercise

The team exercise approach provides little reliability because it is the voice of the team that is being used to rank the requirements rather than the voice of the customer, supplier, or internal operations. However, it is included as an option here because it does force a consensus within the team as to the ranking. Once the requirements are captured, the ranking of the requirements can be tested and measured in an independent survey without diluting the value of the requirement list. If there is great controversy among team members about the rankings, then that is an indicator that the ranking issue must be addressed immediately using an intensive data gathering effort. The range for this column should be from 0.0 to 1.0, where 1.0 is the highest level of importance.

Degree of Supply Chain Satisfaction Provided by Competitors

This data is actually benchmark data. For each requirement, competitors and as well as the company are rated on how well the requirement is being satisfied with current processes if there are any that can be compared. If the roadmap is for a greenfield facility, then any competitors that offer a relevant capability should be included. This rating makes a quantitative statement of the relative strengths and weaknesses among competitors in the market being targeted by using customer, supplier, or internal process satisfaction as the main criterion. This rating allows the display of market benchmarking and competitive information in a simple format so that the entire team can understand the situation. In actual practice, this array of ratings is often one of the most valuable aspects of the QFD matrices because it brings into the clear light of day all the key competitive issues. The ratings in this column should range from 0.0 to 1.0, where 1.0 is for the highest level of satisfaction.

Describe the e-Factory Plan or Vision

In the previous paragraph, the performance of competitors and the company is measured in terms of achieving supply chain satisfaction. In this step, the goals of how much supply chain satisfaction is to be achieved by the e-factory under development is quantitatively articulated. This rating of e-factory goals is done on a requirement-by-requirement basis. It then becomes a clear and unambiguous mapping of the company strategy that evolves into a vision for the e-factory. This is a step that is usually not done in traditional requirements planning methods. In fact, in actual practice, performing this step may create questions that have not been answered by the company strategy, thus requiring that additional work be done on the strategy or that the team must take a position on the strategy in order to complete the project. Either way, there is a benefit because any strategy issues are identified and resolved early in the total process.

All of this contributes to the creation of a vision of the e-factory. The range for this column is from 0.0 to 1.0, where 1.0 for a supply chain requirement implies that the e-factory vision must satisfy this requirement. A score of 0.0 implies that the vision does not include satisfying this requirement. A score of 0.5 implies that the vision does include some form of this requirement. By deciding on the scale for this column, the roadmap team is making a statement about the nature of its vision. If the team decides that partial requirement satisfaction is not acceptable, then the score in this column can only be one of two values, 0.0 and 1.0. If the team decides to have a range of values, this will create a ranking for the requirements a part of the vision.

List Emphasis Levels

An emphasis level factor is a way to give additional weight to specific requirements on top of what has been defined in previous steps. The concept of emphasis is that there may be certain requirements that are more directly related to supply chain success than other requirements. While customers may rank a feature 35th in importance when compared to the other requirements, they may respond to this feature 30 percent more enthusiastically than other features. Therefore, a multiplier of 1.3 would be the emphasis level factor for this feature, while all the other features would be 1.0. This extra weighting would then move the ultimate ranking of this feature up the list from 35.

Calculate the Absolute Rank

This step is simply the multiplication of the Market Importance by the Competitor Satisfaction Ranking by the e-Factory Vision Plan by the Emphasis Level Factor. The total of this column then becomes the denominator for the next step.

Calculate the Normalized Rank

This step is simply the division of each item in the Absolute Rank by the total of the Absolute Rank column. This calculated number now becomes the new rank order for the supply chain requirements that incorporates the e-factory vision, the competitive situation, and the supply chain importance. For most roadmap applications, building the QFD matrix to this point is sufficient to produce a rank-ordered set of supply chain and technology requirements for the roadmap methodology. However, the next few steps are useful to generate even more detail on how requirements should be satisfied.

Create Correlation Values

Creating correlation values between the supply chain requirements and the technical requirements is a somewhat tedious task due to the sheer number of requirement pairs. The job is to identify whether a given technical criterion correlates to a supply chain requirement. Correlation is defined as the degree of relationship between two entities. If technical criteria are extremely important in achieving the satisfaction for a specific supply chain requirement, then it is highly correlated to the customer requirement, and the correlation should be 0.9. If technical criteria are moderately important in achieving the satisfaction for a specific supply chain requirement, then it is only moderately correlated to the customer requirement, and the correlation should be 0.5. If technical criteria have no importance at all in achieving the satisfaction for a specific customer requirement, then it is not correlated to the customer requirement, and the correlation should be 0.0. Because the technical criteria were generated by working down the customer requirement list one by one, there will be at least one high correlation for each technical requirement.

Sum the Technical Requirement Correlation and Normalize

Summing the total of the correlation values of the technical requirements will produce a relative ranking of the technical requirements. The normalized ranking is produced by summing the total of the correlation sums and dividing this into each correlation sum. This row of numbers becomes the normalized forced ranking of the technical requirements.

List Specifications for Technical Requirements

An additional feature of the first QFD matrix is the ability to include numerical specification values for each of the technical requirements generated in an earlier step. This set of specifications becomes the baseline from which evaluation criteria for solution set components can be established.

Summary of QFD Benefits

The QFD approach focuses attention on generating e-factory requirements. The entire objective and purpose of the QFD activity is to have the voice of the supply chain transformed into details that directly drive the e-factory process design. By using the matrix technique, a tremendous amount of information about the requirements and the relationships of the design details to the supply chain requirements can be displayed easily and shared with everyone on the team. The QFD activity begins and ends with attention focused on the three sources of e-factory needs and requirements.

The QFD approach forces the selection of target values for the technical requirements that will ensure supply chain satisfaction. All e-factory technical requirements are chosen on the basis of what is necessary to satisfy supply chain requirements. There is no room in this approach for any individual or functional element in the organization to insert technical requirements that are not related to e-factory requirements.

The QFD approach facilitates the multi-disciplined team approach to product design. The nature of the QFD activity requires that there be contributions from different functions. Marketing and sales provide their input based on their working knowledge of the customers and the market. Engineering provides input on the basis of what has been used in the past and what could be possible in the future. Manufacturing provides input on what the production processes need to be. Strategic planning provides input to (or learns from) the activity because there must be a connection to the e-factory strategy for the matrix to be completed.

The QFD approach documents the decisions made during the e-factory process. Each mark on the matrix documents a decision made about observations, technical alternatives, or specifications. If any decisions that follow come into conflict or contradict previous decisions, then the consequences can be determined by tracing back to the original document.

The QFD approach provides for better communications between functions. Because the QFD matrix is created by a cross-functional team, the entire team is exposed to the same thought process, exploratory discovery of customer requirements, and decision-making about technical approach. While the matrix is being constructed, there will likely be members of the team who will be coming up to speed, so to speak, on the total business process. In the era of specialization, very few, if any, professionals are expected to understand the entire business process of which they happen to be part.

The QFD approach forces collaboration. Collaboration may be too strong a term to describe what happens. The reality is that the QFD activity requires that there be a conclusion drawn on every question that is represented by cells in the matrix. The information entered in the matrix is a documentation of that conclusion. The team participates in the creation of the matrix and in the conclusions drawn. The team environment should be strong enough that given each decision, everyone on the team works together to implement the decisions. This eliminates the time consumed that sometimes occurs in lesser processes by debating issues and second-guessing decisions later in the project. It also forces early involvement. Because the matrix requires input from different functions and because it is created at the initial stage of a project, early involvement in the key enterprise functions is mandated.

Step 3: Creating the Vision

Step 3 is the creation of the vision using the methodology described in Chapter 7. The team continues to work together to build the vision using the steps defined in the methodology.

Step 4: Creating the Roadmap

Step 4 is the creation of the roadmap using the methodology described in Chapter 8. The team continues to work together to build the roadmap using the steps defined in the methodology.

The methodology relies heavily on process scenarios to integrate the requirements and to force the identification of alternatives of how technology and process design can best work together. Scenarios are based on object modeling concepts; it is not sufficient merely to use object orientation. The ultimate goal is to describe what the requirements need to have in the process and technology design.

A scenario is a sequence of transactions in a system initiated by a user of the system. A scenario has a complete flow of events with a well-defined beginning and a well-defined termination.

Scenarios link together different models of a business or an information system. For example, the analysis model can be linked to the design model. They each have an important role in the link between a business model and its information system. When building scenarios, the team must define the workflow of activities for people; the information flow for human, machines, and data resources; the information systems that support the workflow; and the controls for the total system of people and capital.

When the team builds a scenario, it must define the objects; identify which are human, machine, and data objects; and define the control hierarchy level for each object. The team must also define the scenarios by identifying each activity needed to execute the scenario.

Step 5: Creating the Implementation Plan

Creating the implementation plan is the final step for the roadmap team, but is not necessarily the easiest step. Considerable judgment is required in deciding how to approach several of the key questions that must be answered to complete an implementation plan.

For example, one outcome of the roadmap process is the identification of the solution set components in terms of software and hardware suppliers and packages. From this list must come the identification of what must be purchased and when each component should be purchased. This information will have to be put on a project implementation schedule.

Equally important is the definition of the sequence of implementation activities. The roadmap process provides an ordered list of implementation activities. The roadmap team must put specific milestones and dates on that ordered list. This enhanced list of implementation activities then becomes a project schedule.

In addition to this time schedule, the resources that will have to be committed to the project must be identified and the impact of their commitment estimated. If external resources are required, then their cost and time must also be estimated.

After the implementation plan is created, then it must of course be submitted to the steering committee or to management for approval and budgeting.

Summary: Creating a Usable Roadmap

Creating a usable roadmap requires the commitment of a dedicated team of people from a variety of functions from the company, as well as the involvement on a part-time basis of other internal resources, customers, and suppliers. Organizing this team is the first key step in the e-factory roadmap process. It will also likely reveal either new opportunities or issues that the company has with suppliers and customers because it will depend on communication with them directly. The result from the team using the described process should be a roadmap defining the e-factory technology solution set and the business processes.

Chapter 10

Executing the e-Factory Roadmap

Purpose of This Chapter

The purpose of this chapter is to describe the necessary considerations and steps that should be taken to have a successful execution of the e-factory roadmap. Ultimately, the level of success depends on how well the people in the organization and the processes in which they are employed are prepared for the changes and the new technology. The key steps in a successful e-factory implementation include:

1. Making the decision to implement
2. Preparing the business processes for the technology
3. Training or acquiring the people
4. Acquiring the technology
5. Implementing the technology
6. Testing
7. Continuous improvement

Each of these steps are discussed in the following chapter sections.

Making the Decision To Implement

Making the decision to implement an e-factory solution should be a matter for the highest levels of the company. Certainly, the Chief Executive Officer should be not only involved in the decision, but should also be the number-one champion for the program. The Board of Directors should be involved in the decision process as well because such a decision represents a major change

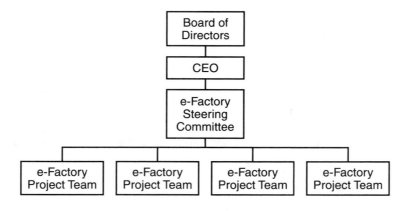

Exhibit 10.1   Corporate Implementation Management Structure

in direction and generally means a major transformation of the company. This type of decision represents one of the most significant decisions that can face a manufacturing company because of the risks if it is not done or if it is not done well. Not only must the potential benefits be understood, but where it fits into the long-term strategy must also be well understood.

Not only should the most senior level of management be involved in the decision, but the management team that is responsible for implementing the decision must also be on same page. Typically, the best approach for managing this type of transformation should have the structure illustrated in Exhibit 10.1.

A steering committee should consist of the senior management team from the various functions or processes that will be affected by the e-factory implementation. The purpose of this committee is to provide the management direction and commitment that is necessary from time-to-time in these projects. The steering committee is also responsible for ensuring that the appropriate resources are available and committed to the various projects and to ensure that the projects are on schedule.

Each e-factory project team should be composed of a team leader and the appropriate people from all the necessary parts of the organization and with all the necessary skills. There will likely be multiple project teams, depending on the pace and breadth of the implementation being undertaken. One of the goals of the methodology described in Chapter 8 is to identify the solution set and produce an implementation roadmap that helps management define how many teams and how many resources are required.

The decision to go or not go should be made by senior management, and the implementation plan should be agreed to by top management and the steering committee. The decision to go should only be made after the planning of the organization and required resource (human, capital, and financial) commitment have been completed. The decision process should consist of the steps illustrated in Exhibit 10.2.

After the decision to implement has been made, then the implementation teams need to be launched.

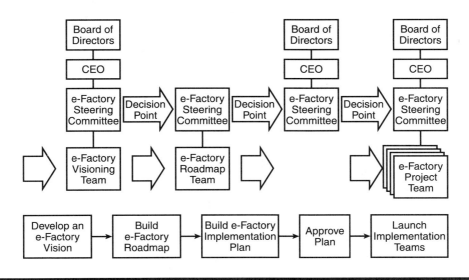

Exhibit 10.2 e-Factory Decision-making Process

Preparing the Business Processes for the Technology

Preparing the business processes for the e-factory technology will require the creation of new job descriptions, new business rules, training documentation, workflow diagrams, and new compensation programs. A flowchart of this process is given in Exhibit 10.3. Even more importantly, there will be a need to build knowledge bases that can be implemented by the technology and that can be used to accelerate the decision-making that will be part of the new business processes.

Much, if not all, of the definition of the new business processes should have been created during the road-mapping effort. The process scenarios that are such a key part of the road-mapping methodology described in Chapter 8 should provide the bulk of the definition of the new business processes. They should have included in them which best process practices are appropriate

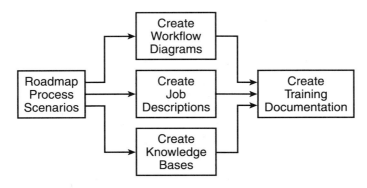

Exhibit 10.3 Preparing the Business Processes for the Technology

for the scenario and which technology will be employed. However, there will likely be a need for further detailed design of the business process by expanding the scenario descriptions to a lower level of detail. More detailed scenario definitions will be required to create workflow diagrams and job descriptions.

Workflow diagrams are necessary for defining the optimum path for the flow of work, responsibility, and information. Such diagrams are useful for defining job descriptions and knowledge bases, and are essential for defining the rules by which the business process operates. Business rules are necessary for defining how the technology to be employed should be configured and which knowledge content should be implemented. Business rules are, in essence, the rule description of the workflow, how decisions are made, and what alternative actions to take under different conditions.

Job descriptions are necessary because they will contain the specific nature of how an old business process is being changed or how a new business process will operate. The job descriptions are necessary as part of the training process and as part of the change management process.

Knowledge bases are the most critical item to be prepared because they represent the company-specific and process-specific content that will be implemented in the technology. Examples of the different types of knowledge bases that typically have to be captured are included in Exhibit 10.4.

Training programs are required to ensure that the ramp-up time for retrained or new employees is as short as possible. Training will be required to teach the employees how to use the new technology and how to operate within the new business processes.

A final important element in preparing the business processes for the new technology is communication. Communication is required before anything happens, as it is happening, and after it happens. It is better to err on the side of over-communication than under-communication. Communication will enable all people to deal with the changes as early as possible and minimize the disruptions that will occur as change happens. While it has not been proven that communication will eliminate the anxieties that people will experience due to change, it does spread it out over time and dilute the intensity so that it is easier to deal with.

Preparing the Employees

No process or technology implementation will be successful unless the people who will be operating the process and technology are prepared. However, before any preparations can be planned, an analysis of the skill sets available in the organization must be performed. From this analysis, an inventory of the human resources and their skill sets can be built. A second analysis to identify the skill sets required to operate the e-factory should then be done. By comparing the current inventory of human resources and their skill sets

Exhibit 10.4   Process Knowledge Bases List

Order Fulfillment Process	Product Realization Process	Customer Relationship Management Process	Support Process	Field Service Process
Bills of materials	Bills of materials	Price list	Financial management archive	Field inventory database
Production process plans	CAD/CAM/ CAE archives	Pricing algorithms	Human resource database	Field personnel demographic database
Production recipes	Product test data archive	Terms and conditions	Intellectual property database	Service manual library
Production equipment process management rules	Product catalogs	Product catalogs	Accounts payable and receivable database	Frequently-asked-questions database
Condition-based maintenance rules	Product and process cost model archives	Product configuration rules	Capital resource database	Service and repair questions and answers
Process control rules	Product simulation model	Customer demographics lists	Training materials database	Service contracts database

to the necessary skill sets, a clear picture of how much recruiting and how much training are required becomes available.

The training program for the e-factory should have three goals: (1) learning how to use the new technology, (2) learning how to execute the new processes, and (3) learning how to make improvements.

Learning how to use the new technology is important because there will inevitably be new terms, new procedures, a new look-and-feel of the user interface, and new questions for each software package and database used as part of the e-factory solution set. These new features may be new because employees had been trained in the legacy systems, or because employees may never have used computer systems before. Although the number of people who have no experience with point-and-click systems is dwindling, many of them can be found in the manufacturing sector.

However, learning how to use the new technology does not ensure a successful implementation of an e-factory solution set. People must be able to work within the new business processes with the new technology. This means that they must learn how to use the new technology to execute the new business process. The ideal situation is to use the business process scenarios, developed during the e-factory roadmap methodology, to define the content for the technology training programs. Thus, rather than having people trained in the generic procedures available from the supplier of the software, they should be trained in how to use the technology the way it is being implemented in the new business processes.

A final goal of training is to provide the employees with an acceptable process for solving problems and identifying improvements for the implemented business processes. No matter how much planning and testing may be done prior to the "go-live" point for the e-factory, there will be a learning and evolution process as people work with the new technology in the real world. To obtain the benefit of this learning curve, there should be a process that permits this learning to be incorporated into the process and the technology.

Preparing the Technology

Preparing the technology consists of acquiring it and then implementing it. Because of the rapidly evolving nature of this field of technology and the evolving practices within it, special care should be taken when planning the acquisition and implementation steps.

Acquiring the technology consists of identifying the team of solution set suppliers (which should have been done as an end result of the roadmapping process), deciding how to manage the team of suppliers with the internal teams, and negotiating the terms and conditions with the suppliers.

The team of suppliers will generally consist of software companies, hardware companies, and often, implementation contractors or consultants. The number and identity of the software and hardware companies depend on the solution set identified as part of the roadmap methodology. The use of implementation contractors or consultants depends on the availability of internal resources and on the availability of time required to get them trained. Typically, there is not enough time and not enough internal resources to avoid the necessity of paying for implementation contractors.

There are several worst practices to be aware of when assembling the solution set team. The biggest risk is that an implementation will cost significantly more money than budgeted and take much more time than scheduled. The next biggest risk is that the implementation will not meet expectations and produce the benefits anticipated.

The cost and time required to implement an e-factory solution set are dependent on many factors. Generally, the definition of a clear set of requirements that covers the realities of the operating environment will minimize cost and schedule risks. The focus on scenarios in the e-factory roadmap methodology is intended to produce just such a set of clear and specific requirements. By forcing the creation of business process scenarios, the specification team is forced to think through all the possible issues that could occur on a day-to-day basis.

Also, the solution set components should have some track record of successful implementations. Unless the e-factory being implemented is intended to be a research project, the solution set components should have been implemented in more than five similar facilities before it is considered a viable candidate. If the e-factory being implemented is intended to be a research project, then higher cost and schedule risks are usually encountered.

There are other factors that add to the risk profiles. If the e-factory solution set components are to be interfaced with or connected to legacy systems, then there may be issues as to how easily and cost effectively this can be done. There may need to be custom programming done to interface to certain older legacy systems. These types of issues should have been included in the roadmap methodology so that the identified solution set will be one that minimizes these types of risks.

Managing the implementation team is also a significant task. There are usually three ways to approach this challenge. The first way is to have a skilled and experienced internal manager be responsible for the entire implementation program. An alternative is to completely outsource this function to an implementation contractor/consultant. A third alternative is to have the steering committee perform the program management function. This third alternative is the least desirable and should be avoided. The first approach is preferable because of the incentive of the internal manager to have a completely successful project. The second alternative of using a consultant is acceptable as long as there are clear performance metrics with which to measure success. An illustration of how an implementation team might be structured is given in Exhibit 10.5.

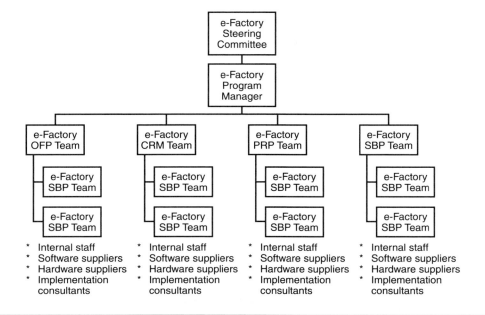

Exhibit 10.5   Implementation Team Structures

Exhibit 10.6   Pricing Structure Examples

Software	Hardware	Consultants
User license fee	Price per equipment	Fee rate per hour
Server license fee	Time usage fees	Fixed fee per project
Corporate license fee	Rental fees	Fee per deliverable
Transaction fee		% of received savings
Fee per Web site hit		
All of the above		

Negotiations with the suppliers should include price, terms and conditions, maintenance and support, and performance metrics. Negotiating price tends to be more art than science because of the wide variety of pricing schemes developed by the suppliers. Examples of the pricing structures available in the marketplace are listed in Exhibit 10.6.

Negotiating terms and conditions is often viewed by corporate customers as an exercise in dealing with minutia. However, it is in this area where the most valuable aspects of the relationship with solution set suppliers can be developed. A list of terms and conditions that are important to an e-factory solution is given Exhibit 10.7.

For major systems implementations, maintenance and support contracts often tend to be overlooked until after the implementation is nearing completion. However, for the e-factory, because of the innovation often designed into the solution and because of the dynamic evolutionary environment that it represents, maintenance and service contracts represent a major part of the total commercial relationship with the solution set providers. A sample list of the key elements necessary in maintenance and service contracts is given in Exhibit 10.8.

Performance metrics are useful in e-factory solutions because they allow management to understand how well the solution is working and also give the solution set suppliers a clear picture of what they need to do to ensure customer satisfaction. A list of example performance metrics is provided in Exhibit 10.9.

Exhibit 10.7   Terms and Conditions Examples

Software	Hardware	Services
Performance levels	Performance levels	Performance levels
Uptime levels	Uptime levels	Intellectual property
Intellectual property ownership	Upgrades and bug fixes	ownership
Upgrades and bug fixes	Dispute resolution	Training support
Dispute resolution	Training support	Implementation
Implementation schedule	Installation schedule	schedule
Training support	Documentation support	

Exhibit 10.8   Maintenance and Service Feature Examples

Software	*Hardware*	*Services*
Upgrades and bug fixes	Upgrades and bug fixes	Training support
Rapid response service (24 hr)	Rapid response service (24 hr)	Process audits
Training support	Training support	On-line support
Remote diagnostics	Remote diagnostics	Documentation
Condition-based maintenance	Condition-based maintenance	support
Off-site disaster recovery	Spares management	
	Off-site disaster recovery	

After all the commercial issues have been resolved and there are signed agreements with all the suppliers of the e-factory solution set, all parties involved can then get down to implementing the technology. Implementing the technology requires that several topics be addressed.

The first topic is where the implementation activities should be performed. Implementation location refers to where the implementation team members reside; where the hardware being acquired resides prior to, during, and after implementation; where the software is hosted; and where the testing is done. There are several options for locations where the implementation might be performed. One option for location is at the e-factory site itself. Another option is to set up a special implementation laboratory facility separate from the actual e-factory site where all the technology and people can work together without being distracted by ongoing business operations. A third option is to have a virtual location, where the individual elements of the technology team reside in their normal locations and the interaction among the team members is done via electronic collaboration. A fourth option is some combination of the above. For example, a project could start as a virtual location, move to a systems integration laboratory to complete design and test, and then end up with an on-site installation.

Once the location for the implementation team members has been decided, then the team can get down to implementing the technology. The implementation effort ususally consists of configuring the systems, implementing content and building databases, and designing a fail-safe backup and recovery capability.

Exhibit 10.9   Performance Metrics Examples

Software	*Hardware*	*Services*
Cycle time for specific processes	Transaction cycle times	Schedule conformance
Data security	Uptime and availability	Requirement conformance
Number of simultaneous users	Storage capacity	Special project response time
Fail-safe operations	Processing speed	Staff availability
Data integrity	Security	
Cost per transactions	Number of simultaneous users	

Configuring the system is significant because of the variety of options typically available from the hardware and software suppliers. Configuring the systems usually means implementing the business rules that have been defined during the scenario building process and detailed during the process preparation activities, selecting the various technology options available from the software and hardware suppliers, and implementing whatever special or custom interfaces that are required.

Of course, the technology solution is irrelevant without the knowledge content that is necessary to run the e-factory and support the interactions and decision-making process. Because the e-factory is a real-time environment, content is at an absolute premium. Real-time technology environments require that all content necessary to execute the decisions and the transactions called for by the real-time environment must be built into the system. Content means databases, decision-making rules, and integration with data capture technology.

A third key part of a successful implementation is the creation of a viable backup or fail-safe capability for the e-factory solution set. The real-time environment means that it has to be impervious to any technical glitches or failures to individual solution set components. The definition of the requirements for this capability should be captured during the roadmap process; thus, the solution should be well-defined prior to the start of implementation.

Testing

Testing the implementation of the e-factory solution set is probably the most important activity for which there is typically not enough time scheduled or not enough detailed planning. In theory, if the implementation is well-planned and process controls are in place during the implementation, then testing could be minimized. However, technology implementations have not generally been characterized by first-pass yields of 100 percent. Therefore, testing is a necessary and sufficient task for any e-factory implementation.

There are only a few key steps in a successful test plan. These steps include (1) developing a test plan, (2) testing the processes, (3) testing the technologies, and (4) evaluating the results.

Developing the test plan for the total e-factory solution set involves integrating the test plans for each of the key components, integrating the test procedures of each of the solution set suppliers, creating test plans for the integration activities not covered by supplier or component plans, and ultimately managing the interaction and schedules of all the parties involved in the team efforts.

Testing the processes should be based on using the business process scenarios developed during the roadmap activity to define the test scripts and procedures. The purpose of testing the processes is to ensure that the technology implemented supports and executes the functions that are needed by the process, that the content implemented supports the processes appropriately, and that the process as implemented performs as defined by the scenarios.

Testing the technology should be based on achieving the performance specified by the requirements developed during the roadmap process. Usually, the key challenges are to ensure that the latest features of each software package operate as promised, that the different software packages interact with each other as efficiently as possible, that the databases and content implement operate as expected, and that the overall performance is achieved.

It is important to not overlook the value of evaluating test results. Tests should not be designed to be pass/fail. If a test is carefully designed, there can be information about the quality of the process design as well as about the performance of the design. If a test is designed to be pass/fail, then information is available only when a test fails. If the test is passed, the only information available is that the process did not fail. If the test is failed, then it is clear that something must be fixed. A poorly designed process can pass a poorly designed test. A process that is operating very close to a failure mode can pass a test. An intelligently designed test will provide performance statistics and other measurements that can be used to indicate trends or how close the process might be to failing.

Beyond Implementation: Continuous Improvement

The creation of the e-factory does not stop the moment that implementation is complete and the "go live" signal is raised. The e-factory environment is so dynamic and evolutionary that there must be a process for feedback of ideas on how to improve the environment and how to change the environment to respond to changing enterprise needs. During the past two decades, the general principles of best practices related to achieving a "continuous improvement" culture in business organizations were fully developed under the banner of Total Quality. However, these same principles are applicable to an e-commerce environment. In fact, these principles may be even more effective in an e-commerce environment because they provide for a process by which ideas for improvement and change can be rapidly identified and incorporated into the existing process.

Continuous improvement methods are well-established. They consist of a standard set of problem-solving tools, standard terminology for quality and process improvement, and periodic (weekly/daily) improvement meetings in which people can discuss the status of the current situation and identify opportunities for improvement.

Creating a process for identifying the opportunities for improvement will rapidly evaporate unless there is also a process for implementing the improvements once they are identified. This improvement process must support technology improvements as well as process improvements.

Summary

The roadmap methodology facilitates the identification of the best solution set for a given situation. Implementation of the solution set requires a significant management effort to harness all the resources of the solution set suppliers, consultants, and internal staff. Ultimately, launching the e-factory implementation effort launches a transformation of the manufacturing enterprise that cannot be turned back. The decision to launch must be part of an overall company strategy that has as its goals the fundamental transformation of how it adds value to its supply chain and of how it operates within the transformed supply chain.

Chapter 11

Final Comments and Looking Forward

Purpose of This Chapter

After reviewing the trends in creating the e-factory environment and examining a methodology for developing a roadmap on how to achieve an e-factory, this final chapter provides thoughts on where the course of future developments in both process and technology might lead. While the future is usually built on what has happened in the past, the future in the e-factory will also be an opportunity for creativity and innovation that has not existed for decades.

Evolving Best Practices in the e-Factory

There has been an evolution of business process practices during the past ten years that has resulted in a set of principles and concepts that have merged from a variety of sources. All of these principles were created with the concept of having the most efficient process in terms of resources consumed, the fastest process in terms of cycle time, and of having the highest quality process in terms of eliminating mistakes. The advent of e-commerce has added flexibility as an additional requirement to the emerging set of best "e-practices" in the business processes for the e-factory. All of these principles are important because they either create or facilitate the benefits of operation that e-factory technology promises.

The first major breakthrough in the evolution of business process practices was in process streamlining. Process streamlining is simply the elimination of non-value-adding activities and the execution of activities in parallel where possible. Process streamlining has evolved from the fusion of process techniques such as JIT, TQM, ABM, and 6-Sigma. There has evolved a consensus of best practices from these techniques that all have the same objectives: elimination of wasted effort and wasted time. Prior to the emergence of e-commerce and e-factory requirements, the elimination of wasted effort and time could be achieved with careful design of workflow. With the e-factory technologies available, new opportunities for workflow design and work acceleration exist so that even greater performance improvements can be achieved.

Flexibility in business processes is now more necessary and more practical due to innovation in thinking about how processes should be designed, as well as in the technology that can be deployed in implementing processes. Flexibility comes primarily from rapid reconfiguration of workflow and material flow. Rapid reconfiguration of workflow requires that any changeovers in equipment setup be performed as rapidly as possible, and that any changes in control systems occur as rapidly as possible. Rapid configuration in material flow requires that material handling and operating instructions on the factory floor be changed as rapidly as possible. For this flexibility to occur, the design of the processes, the setups, the information systems, and other manufacturing technologies should be as integrated as possible.

Process quality has the opportunity to reach new heights because of the availability of new control technologies integrated with workflow. The simple operating principle behind good process quality is to eliminate variation. If a process is designed to produce a perfect product, the only cause for a mistake is if there is a chance for there to be a variation in the process. Variation occurs when an activity is not completely controllable. If a human is performing the activity, then weariness, distraction, poor training, and poor feedback are sources of variation. If a machine is performing the activity, then loss of machine tolerance due to wear, poor feedback of operation, random events in the environment, and poor material setup or fixturing can contribute to variation. New quality and process control technologies make it possible to eliminate many of these sources of variation, thus making it possible to achieve unprecedented levels of quality. This opportunity will put new pressure on design processes.

Best Practices in the Product Realization Process (PRP)

During the past ten years, there have been several new practices (and some not so new) that have emerged as providing the best approach in the product realization process (PRP). These "new" best practices include:

- Process stages that have go/no-go decision points
- Multi-dimensional (cross-functional, cross-organizational) teams that own and execute the product realization process
- Enabling tools that improve and accelerate decision-making
- Collaborative design philosophies that design the supply chain and the product simultaneously
- Product strategies that integrate product, supply chain, and market needs

These new concepts are less concepts and new philosophies than they are fresh views of how to determine what is important in product realization and how to use the best analysis and synthesis approaches to design products.

A product realization process should consist of stages and go/no-go decision milestones. Each stage is performed by a collaborative "team" that is cross-functional. At the end of each stage, a go/no-go decision on advancing to the next stage is made.

- There is usually a "deliverable" at each stage, such as:
  — Business plan
  — Functional specification
  — Design specification
  — Production prototype and documentation
  — Final model and documentation
- Stages often include:
  — Capture "voice of customer"
  — Develop conceptual designs
  — Finalize detail designs
  — Test prototypes
  — Get ready for production
  — Launch production

A diagram of generic stages is given in Exhibit 11.1. The process should allow for multiple project paths. If a project is determined to be an extension of an existing product family, then one or more stages can be skipped or minimized. If a project requires innovation and new product concepts, then the full set of stages will be required.

A PRP should be performed by a multi-dimensional (cross-functional, cross-organizational) team that owns and executes the product realization process.

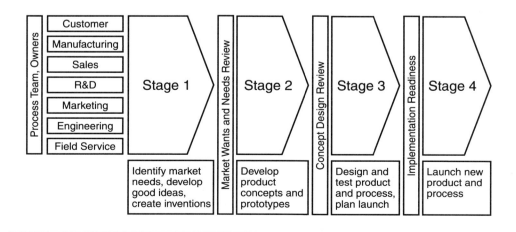

Exhibit 11.1   Example of a PRP with Stages and Decision Milestones

Cross-functional and cross-organizational teams are essential to executing a PRP that is rapid and that satisfies the voice of the customer. Such teams are also necessary to ensure that simultaneity of the design of the production process, the supply chain, and the product is achieved. The teams must have clear charters and access to a steering committee or sponsoring group that has the authority to make decisions and eliminate internal barriers.

A PRP should use enabling tools to improve and accelerate decision-making (see Exhibit 11.2). Working harder to speed up the execution of an activity may work in special situations, but as a general form of improvement, it is difficult to institutionalize. During the past few years, there has been the realization of Web-enabled "enabling tools" that can reduce the time and thus accelerating many of the decisions that are part of the key activities in a product realization process. These enabling tools allow the product realization team to work intelligently and collaboratively.

The PRP should foster the simultaneous design of the supply chain process, the e-factory, and the product. Designing the product is not enough for rapid product realization. The e-factory processes to be used to build the product and the supply chain that supports the product must be consciously designed at the same time. If the new product is being designed to use an existing production

Exhibit 11.2 Enabling Tool Matrix

	Product Performance	Cost Minimization	Robust Design	Customer Satisfaction	Time-to-Market
Quality Function Deployment (QFD)	X	X	X	X	X
Invention Knowledge Base (IDB)	X			X	X
Design of Experiments (DOE)	X				
Product Cost Modeling (PCM)		X	X	X	
Product Data Management (PDM)	X	X	X	X	X
Design for manufacturing (DFM)	X	X			X
Failure Mode Effects Analysis (FMEA)	X		X	X	
Computer Aided Engineering (CAE)	X	X	X	X	X
Product Configuration Tool (PCT)				X	X
Design for other criteria (DFX)	X	X	X		
Value Engineering (VE)	X	X		X	

process or supply chain, then the rules and constraints from those existing processes must be used for decision-making about the new product design.

The PRP should ensure an integrated strategy for product, supply chain, and markets. Virtually all of the key decisions made about the new product as it moves through the PRP stages relate explicitly or implicitly to the product strategy, and the product strategy should be driven by or connected to the market and business strategies. Too many companies operate with an open loop between their general business strategies and the product strategy that is operating within their units.

Best Practices in Order Fulfillment Business Processes

The most significant recent trend in order fulfillment is that speed is the most important parameter — not resource utilization. So much of what technology was asked to do and what strategy intended to do was based on maximizing the utilization of human and capital resources. This is still important because any resource unnecessarily under-utilized is a wasted resource; but when process design is viewed from the perspective of maximizing speed, the decisions on what resources to deploy and how to employ them change.

One of the best practices is to design business processes to be continuous flow cells. Cellular design forces the process designer to focus on breaking down an entire production process into groups of activities that maximize throughput, minimize inventory, and provide flexibility. Viewed another way, cellular design is based on the use of modularity concepts applied to process rather than product. Cellular design may require a trade-off between equipment utilization and speed, but speed should win in most cases.

Another best practice is to design business processes with a strategy of build-to-order rather than build-to-forecast. Designing a business process with a build-to-order approach means that the size of lots flowing through a line of cells will be small (possibly even of size one), that the variety of products that should be manufacturable through a line of cells will be broad, and that the flexibility of the machines and the work team will be maximized.

Many of the principles described above come from the concepts developed under just-in-time (JIT) philosophies. The basic point of JIT is to provide something only when it is needed. This philosophy forces the design of the business processes to be streamlined, fast, and clean.

A key element of these best practices is having the ability to rapidly reconfigure workflow, material flow, people, and machines. This can only be achieved by designing the machines to have flexible setups mechanically and electronically, for the people in the process to be trained to operate in flexible teams, and for all the resources to be as mobile as possible.

Ultimately, the degree of flexibility and speed will depend on technology. Technology is critical to the acceleration of work activity, the integration of information, the control of quality, and the reconfigurability of work and material flow. In the e-factory, technology is used to facilitate, execute, and accelerate the application of the best process practices.

Best Practices in Customer Relationship Management Business Processes

Customer relationship management is a new field that was created almost entirely by the trends in Internet-based e-commerce. The sum total of what customer relationship management is about is that the entire procurement and purchasing experience for a person or a company is on-line and real-time. This phenomenon by itself has major implications for how a company designs its products and processes.

First of all, for the shopping and buying experience to be on-line, all information necessary for that experience needs to be on-line. This means capturing not only all the product catalog information, all the advertising and brand awareness information, and pricing information, but also all the level of service and trust in relationship information that used to exist only between people such as salesman and buyer or store clerk and consumer. It also means capturing and representing information on how to use the product and what to buy, as well price and availability. It also means providing more comparative information on competing alternatives.

However, the trends do not stop at just converting the shopping experience from tactile to digital. There will need to be more on-line collaboration between the potential customer and the potential supplier. Entire new services will be spawned by the need to provide either simplification in the on-line experience (e.g., aggregators) or more interactiveness and content in the on-line experience (e.g., e-negotiation) as to what is needed in the future. There will need to be more support for negotiation of terms and conditions — not just prices, for example.

Best Practices in Supply Chain Management Business Processes

The impact of Internet-driven trends may have the biggest impact on supply chain management because of the major changes in thinking about what form relationships between companies in a supply chain should take and even about what the structure of companies in a supply chain should be.

Internet trends are leading to the need for and the opportunity for more collaboration in planning, scheduling, execution, settlement, new product development, and external processes. Collaboration implies that companies share information and intellectual property to extremes that would not have been conceivable prior to the Internet. Collaboration also implies that there will be more flexibility in constructing commercial relationships and contracts. In the early stages of this trend, collaboration is taking place in an experimental fashion as different approaches to "sharing" are taken. However, as companies learn from these experiments, best practices and rules on how collaboration should work are emerging.

Another key trend is the occurrence of more structural changes as companies specialize on focused processes, as they decentralize, and as they send more activities to outsourcing. The ultimate outcome is the creation of supply networks in which companies become value-adding nodes within the network. Each node could be adding value as a manufacturing cell, as an engineering team, or as a source of services.

As supply chains become networks or webs, there will be a need to focus on rapid reconfiguration of the relationships between companies within a supply network to create virtual supply chains. For this reconfiguration capability to exist, there needs to be new thinking about how decisions are made between company nodes in a supply network, how collaboration could be performed, and how partnering between companies occurs. These capabilities are beginning to emerge as part of new e-market portals and new Web-enabled negotiation technologies.

In addition to the capability for more creative decision in a supply web environment, there is also emerging more distribution of responsibility for inventory management to nodes in the supply web that are remote from the source or owner of the inventory.

Best Practices in Support Processes

There has been a consistent trend and proven cost benefit of outsourcing of key functions such as payroll, human resource data management, benefits management, training, and personnel evaluations. In an Internet-based economy, not only is the outsourcing of support process functions a best practice but having these services on-line is rapidly becoming a best practice.

Payroll services are on-line, or at least available on a remote basis from financial institutions and payroll service firms. In the e-factory, payroll services should be provided by computer-based attendance and timekeeping systems for hourly workers in combination with electronic payroll programs. These capabilities can be installed and owned by companies as part of their enterprise system, or they can be accessed from a payroll service provider on-line over the Internet.

Human resource data management has traditionally been performed with an in-house system. Most ERP systems include this capability in their offerings. However, as more companies outsource their human resource services, the HR data management function is becoming an on-line, real-time system available through the Internet.

Benefits management and communication have been moving faster to the Internet than most every other supporting business process. One of the key advantages of having all benefits information on-line is that it provides instant access to all employees at any time of the day or night and frees the HR professional staff from having to provide this service. In the e-factory, benefits management and communication is definitely on-line.

On-line, Web-based training systems are becoming more popular for organizations that have to change quickly and that are far-flung. Computer-based training and education have been a dream of productivity improvement-minded educators for decades. Combined with the infinite reach of the Internet, Web-based training is a best practice that is emerging slowly but surely in the world of e-commerce and the e-factory. In the future, the e-factory will include on-line training accessible from local training rooms, on computers at the workcenter, or on the factory equipment.

Financial reports for all types of operations are available from ERP systems today. The nature of the reports available and who has access to them are a function of the configuration of the system implemented. In the activity-based, cost-modeled e-factory, financial reports on everything from the units being produced to the profitability of the machine, cell, and production line will be available at the machine, cell, or line. As the e-factory becomes a node in a supply network of nodes, financial reports on the operation of a node, cell, or machine will be important to the local operators and to the entity managing the network of nodes.

Best Practices in Field Service Processes

New service offerings based on on-line transactions and on-line capture of machine condition data for customer support, maintenance, and installation are rapidly evolving. By having intelligent devices controlling and monitoring the condition of equipment and the process that the equipment supports, the readiness and service needs of equipment can be analyzed by remote diagnostic and monitoring software over great distances.

Not only can maintenance services be managed over the Internet, but spares inventory can also be managed. In addition, the dispatch of all human and capital resources can be managed on-line. In essence, all the capabilities for real-time maintenance to be found in an e-factory can also be implemented for a distributed network of e-factories. Thus, not only can the maintenance and care of equipment be outsourced, but it can also be put on-line.

These capabilities are not only becoming an essential ingredient of field service organizations, but they are also becoming part of the design requirements of new equipment. The e-factory will be populated by manufacturing equipment and machines of every type that have designed into them remote software that will perform self-diagnostics, communication technology that will all diagnostics to be performed remotely, and enough processing horsepower to allow condition-based maintenance algorithms and predictive maintenance analysis to be performed.

Best Practices in Technology

During the rush to get information systems ready for the Year 2000 bug, there were many debates among consultants and experts about whether the best approach to information technology was to use a single vendor-provided enterprise system or a multi-vendor best-of-breed solution set. Because the

Internet has become the key driver in economic decision-making, the best-of-breed approach has become the preferred approach to satisfying the information technology needs of the modern corporation. The numerous mergers and acquisitions and joint marketing agreements between enterprise software vendors and special-purpose software vendors have been clear indicators of this trend.

Web-enabled systems make it more practical for the best-of-breed approach to be successful. The hierarchical centralized structure of ERP systems does not lend itself to intelligent machines, cells, and nodes operating in a supply chain network. In the new era of best-of-breed, Web-enabled, and networked nodes, database management is the key to success. As the functionality to support or replace key activities in business processes becomes available in a variety of software packages, it becomes clearer that the process content usually found in the form of databases becomes the unifying common denominator in executing the functions necessary in the e-factory. There are many options for presenting information on-line; it is the content of databases that will make the experience worthwhile. The best-of-breed approach makes it even more important to have a roadmap for defining and implementing the technology and the best processes for the e-factory.

What's after B-to-B and the Impact on the e-Factory

Because the e-commerce era is still so new and the variety of terms coined has been growing exponentially, it is difficult to create a chart that illustrates all the changes that have been occurring in the world of the Internet. The diagram in Exhibit 11.3 is an illustration of three waves of Internet-based evolution. The first wave captures most of the early business-to-consumer (B-to-C) capabilities. The first wave started with companies putting up Web sites that posted product catalogs or advertisements. Later in this wave, retail transactions could be performed on-line. The second wave started with business-to-business (B-to-B) sharing of information and evolved to on-line transactions. The third wave, which is now underway, goes beyond the one-dimensional B-to-B and includes the multiple dimensions of collaboration, interactive internal and external business processes, and dissemination of Web-enablement, down to smaller and smaller components of the enterprise. The e-factory is in the third wave of the Internet revolution. And because the third wave has just begun, the entire picture of the future is not totally clear.

e-Collaboration

One of the fundamental tenets of e-commerce and the e-factory is being on-line and real-time. The immediacy and virtuality of this environment makes it possible for people to interact with each other from anywhere, at anytime. This creates the obvious opportunity for people to work together to accomplish

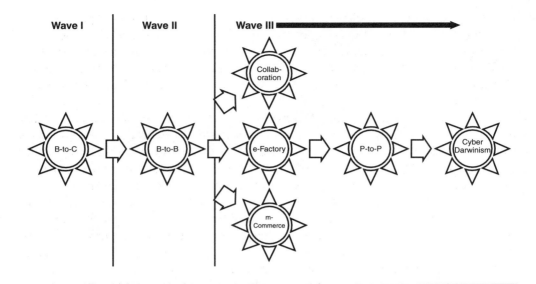

Exhibit 11.3   Next Wave Possibilities

an objective, to complete a project, and to produce a result. However, to go beyond just a chat room interaction and achieve a true collaboration requires something more than just immediate access. There needs to be an electronic environment in which people can interact with each other and with knowledge bases; in which ideas can flow freely yet there be a mechanism for achieving order and decision; and in which people can negotiate and resolve conflicts efficiently.

There is a significant history of experience amassed by software companies and individuals offering collaboration capabilities. These capabilities include features such as:

- Organizing information by project
- Publishing information on the Web
- Providing password protection
- Supporting meetings, discussion, and documentation:
  — Meetings support that includes record minutes, action items, and notes editing
  — Discussion support that includes exchange messages related to projects in a single place
  — Document support that includes store and exchange documents related to a project
- Administering collaboration (which includes the addition and deletion of users, meetings, discussions or documents)

However, there has been collaboration software available for many years, usually going under the banner of "groupware." More is needed for it to become a major new wave of Internet innovation.

e-Negotiation

Getting buyers and sellers together electronically is the heart of much of the current wave of e-commerce and B-to-B e-procurement capabilities. Most forms of market-making today are either an auction where the only negotiation is on price, or an electronic purchase from a product catalog where there is no negotiation. New technology is being developed to support interaction between parties that most resolve specific issues that go beyond price. Topics that will soon be amenable to on-line negotiation include:

- Price
- Product requirements
- Delivery requirements
- Warranty terms
- Payment terms
- Financing alternatives
- Intellectual property ownership
- Testing and validation requirements

Negotiation engines are beginning to appear on the Web. A negotiation engine should orchestrate complex, information-intensive agreements by using dialogue-based transactions between the parties involved in a transaction. In many ways, a negotiation is nothing more than assembling a legal agreement from a catalog of contract clauses, price drivers, and financial arrangements. The negotiation then becomes an interactive two-player game that results in a compromise rather than total supreme victory as in popular video games. While negotiation technology is well within reach, the content and structure of the negotiable components requires much work. As companies achieve success in building their on-line product databases, they will next need to build terms and condition clause databases.

P-to-P

The logical next step to business-to-business (B-to-B) is process-to-process (P-to-P). Whereas B-to-B refers to the interaction and transactions between two business entities, P-to-P refers to the interaction between two or more business processes that exist within the same or different business entities. While this could be viewed as a subset of the current B-to-B phenomenon, as it evolves to more extreme situations, it will create an entirely new dynamic. In most companies, there is a huge amount of cost and value locked up in the internal business processes, and any opportunities to release the value and eliminate the cost will be rapidly employed.

This is already occurring in the form of the external business processes discussed in Chapter 4. External business processes represent the *de facto* or implicit set of activities performed between two enterprises so that a transaction or exchange can be achieved. Today, these external business processes are taking place during B-to-B relationships such as procurement–supplying, buying–selling, specifying–developing, and invoicing–paying. It is clear that B-to-B is, in reality, a set of P-to-P relationships. However, today's B-to-B relationship is executed through only one or a small number of Web-based portals or through EDI.

In the future as e-factory solution sets evolve, each business process or subprocess could be controlled by a Web-enabled solution set component. This component could be any one of the solution set components discussed, for example, a server-based factory floor control system or an Internet server-based procurement system, or some other form that does not yet exist. If the business process has been designed using best practices, it is organized as a cell. The enterprise then becomes a network of Web-enabled process cells. If the e-factory solution Web-enables each business process, then the entire e-factory enterprise becomes Internet based, and the entire enterprise becomes a nodal network of software-controlled process cells integrated by the Internet. The physical enterprise thus becomes the virtual enterprise.

Adaptive Supply Chains

If the concept of P-to-P (process-to-process) is taken a step further to include what has traditionally been called the supply chain, the possibility of assembling supply chains from process nodes becomes viable. Being able to assemble and reassemble supply networks from process nodes and e-factories leads to the concept of adaptive supply chains. Adaptive supply chains imply flexibility, quick changes, and reaction to external forces in a way to optimize results.

Adaptation may be required to respond to fluctuating market demand. If a new product starts out with low-volume sales but then becomes popular in the marketplace, there would be a need to switch to high-volume production. This could be done by rapidly changing the supply chain to include more e-factory nodes or higher-capacity e-factory nodes.

Adaptation may be required to respond to changing market needs. If new product opportunities arise, then the supply chain should change to include new product design nodes and new e-factory nodes.

In some ways, the implications of the third wave that include adaptive supply chains are a throwback to the supply webs of Renaissance Europe. During the Middle Ages in Europe, craftsmen and tradesmen organized guilds that were flexible supply chains of people who could work individually or in teams to accomplish small and large projects. While the term "Renaissance person" refers to one who is broadly capable and multi-talented, Northern Europe did not begin to flourish until webs of craftsmen and central markets began to organize. The third wave of e-business will be based on a dynamic environment of rapidly organizing and changing supply webs.

However, for agile and adaptive chains to work, some type of controlling mechanism is required. This mechanism should provide for a supply chain organizer that provides or facilitates the organization of a supply network which allows value to be added efficiently, and that allows a financial return on the investment to be made in each node. This supply chain organizer could be based on negotiation, search engines, pricing models, and intelligent agents.

Much of the early work in the area of adaptive supply chains has been sponsored by the Department of Defense Advanced Projects Agency (ARPA), the same organization that sponsored the original development of Internet technology. In the early 1990s after the fall of the Soviet Union and after the defense industry began to shrink, the U.S. government became concerned about the ability of the U.S. economy to convert to building military goods if a buildup was required due to some global crisis in the future. This concern created interest in how commercial supply chains in the U.S. economy could be incentivized and adapted to build military equipment.

ARPA and the National Institute of Standards and Technology (NIST) have funded programs to develop innovative new solutions to problems in scheduling and controlling factories and supply chains using software agent technology and complexity theory. This work is based on the theory that large numbers of relatively small or "fine-grained" agents, each operating autonomously within a very small area of the problem environment, can be used to control very complex systems. Autonomous agents have analogs in nature (e.g., ant colonies or beehives). The colonies and hives consist of thousands of individuals that have a few simple behaviors but as a group can accomplish significant tasks without any centralized control. Similarly, in the e-factory or in the supply chain, numerous small agents representing individual cells, machines, parts, operators, and other nodes can operate successfully and under control despite the complexity of scheduling an interactive environment.

e-Markets

The rapid evolution of e-commerce for B-to-B to get all buyers and sellers on-line is leading to the creation of "e-markets." An e-market generally refers to an industry-specific site that is a virtual market where buyers find sellers, and vice versa. The concept of e-markets gained quick acceptance in a variety of industries during the early part of the year 2000, but then diminished toward the end of the year. The promised benefits of an e-market or market exchange were based on cutting through the slow and inefficient traditional system of sales visits and trade shows where business is conducted based on long-term relationships between sales representatives and purchasers. Forecasters predicted that significant percentages of all trade would flow through e-markets within the next five years.

However, e-markets are replacements of physical markets and are not necessarily transformational. Like much of the rapidly expanding world of B-to-B e-commerce, e-marketplaces are driven by procurement. Corporate purchasers are eager to take the savings that come when their vendors find themselves in direct and open competition. Vendors respond quickly to the new sales environment in order to protect their market.

Industries where e-market trends could grow include computing and electronics, shipping and warehousing, and utilities industries. Heavy industries, aerospace, and defense may find less of their commercial transactions flowing through e-marketplaces because such networks favor mass products that can be sold openly, rather than products that need to be negotiated, designed, and manufactured to precise specifications.

In time, even the market for customized products could move to e-marketplaces. IBM's recent launch of Component Knowledge, for example, is a model for network purchasing of electronic components for original equipment manufacturers that allows the purchaser to communicate with multiple component manufacturers in the process of procuring custom-designed components. The e-negotiation tools described above could be coupled with e-markets to serve the custom product transactions.

There were dozens of e-markets announced in the year 2000 alone. Whether any of these e-markets survive remains to be seen. However, if history is used as a guide, the failed e-market experiments will lead to something more useful and beneficial as a new form of e-commerce.

m-Commerce

If every person has a mobile personal communication system (i.e., a cell phone or a palm computer), then e-commerce will be done in a mobile environment. Rapidly becoming known as m-commerce, the early emergence of this wave of on-line business is occurring first outside the United States. A service called "i-mode" introduced in 1999 by Japan's NTT DoCoMo, Inc., the wireless arm of Nippon Telegraph and Telephone Co., provides subscribers with the ability to shop on-line with their joystick-equipped cell phone and receive one bill at the end of the month for all the transactions. In Europe, consumers can pay for parking, wine, and meals using their cellphone-based on-line services. In the United States, several cell phone companies are beginning to offer the ability of a limited number of products online, but the rate of introduction of expanded service trails that found in Europe and Asia.

While the introduction of m-commerce is occurring first in the B-to-C portion of the supply chain, it will likely be introduced rapidly to the B-to-B applications in the future. The benefit in the industrial environment is greater flexibility and lower cost by avoiding the need to invest in wires and networking equipment in the plant itself. The typical cost of a high-speed local area network is measured in hundreds of dollars per wall outlet. The cost of palm PCS technology is competitive with this cost and provides for much greater flexibility.

Other implications go behind human communication to machine-to-machine communication and human-to-machine communication. Because each machine in the e-factory is intelligent and can communicate, the simple step is to include a cell phone chip or board in a machine tool rather than an Ethernet card. Then the e-factory becomes wireless, making it easier to move equipment and people.

Aggregators

One of the barriers to an even more rapid growth in the number of commercial transactions being performed on the Internet is the amount of time it takes to do the Web-surfing to go from Web-store site to Web-store site to do the shopping necessary to make a successful buying experience. For many people, the time it takes to shop on-line rivals the amount of time it could take to shop in the local mall. In addition, there are a lot of Web addresses, user IDs, and passwords that are necessary for the secure on-line experience. While it was only three or four years ago that several companies began trying to offer a broad shopping experience with a Web mall, many of these experiments have ended in expensive failures.

However, a new type of solution is emerging called account aggregators. A wide variety of companies are beginning to see their B-to-C future as one in which they have to offer one Web site where a consumer can view and track all his or her Web accounts using a single log-on, user ID, and password. This offers extreme convenience and significantly reduces the on-line time. The aggregator sites use a practice called screen scraping to grab the user interfaces at all the Web sites for which a user wishes to aggregate. The Web portal and financial institutions began the surge, but it is being picked up in a variety of other industries. One of the factors accelerating this trend is the desire by every company to keep other companies from coming between them and their customers. This CRM battle is being fought in financial services, healthcare, distribution, and manufacturing industries.

Consumer account aggregation is rapidly expanding to the B-to-B industrial account aggregation. While the consumer benefits from having to go to only one Web portal to find news, e-mail, make purchases, make reservations, and browse for other services, the industrial customer can benefit from the aggregation of commercial relationships. The threat of disintermediation that began in the consumer goods and services industries is now appearing in the industrial distribution industries. If a buyer of commodities can go to only one Web portal to shop for the best buy in materials or catalog parts, that saves time and provides for easier comparison shopping. The industrial distributor has traditionally served the role of one-stop shopping and delivery of a wide variety of products and services. This was accomplished by investing in a large physical network of stores and inventory. If the suppliers of the products and services have e-factory capabilities and response times and the shopping can be done on a single Web site, an industrial account aggregator can become a virtual distributor without investing in retail space or inventory.

Cyber Darwinism

Although the debate about whether evolution is a valid theory for the creation of mankind has been going on for two centuries, the concept of evolution is gaining ground as a philosophical approach to the design and development of new products and technologies. The traditional philosophical approach to product development has been based more or less on the scientific: hypothesize a concept, develop the concept, test the concept, and if it fails, repeat the process. The evolutionary approach is based loosely on the theory of natural selection: that only those random changes in a current design will survive if those changes create a better or more survivable design. Taken to its extreme, the evolutionary approach to product design is more like a brute-force acceleration of evolution: create an exhaustive number of random perturbations to a design concept, challenge all of the designs with a do-or-die test, and keep only those that perform the best.

This approach is aready gaining acceptance by some industries. Combinatorial chemistry is now being used in the pharmaceutical industrial as a method for designing new drugs. The basic approach in combinatorial chemistry begins with the definition of a desired chemical reaction, followed by selection of a set of initial chemical structures, creating all possible combinations of variations to the chemical structure within certain constraints, and testing all the combinations for producing the target reaction. If a structure survives the test, then that structure becomes the initial structure for the next phase of the search for a chemical reaction.

One of the most important tools in evolutionary design methods is computer simulation. In combinatorial chemistry, for example, most of the testing of chemical structure variation is done through computer simulations. Computer simulations speed up the testing process because the actual physical testing of chemical compounds would take time and money. In fact, some of the chemical compounds may even be impossible to create; ideally the computer simulations would be able to identify the impossible. Of course, there must be enough theoretical knowledge about chemical structures to be able to create the software that simulates the physical environment being tested.

Thus, one of the key ingredients to evolutionary product design is the availability of computer software for simulating the environment within which the new products will have to operate. This is an area of simulation and knowledge modeling that is an offshoot of the research and development into artificial intelligence during the past two decades. However, rather than trying to model human reasoning as artificial intelligence was, evolutionary product design is trying to model natural selection, which is possibly a more reachable goal.

For many industries, this type of technology is beginning to emerge. In fact, this is an area of research in several universities and corporations. One of the areas of research possibly inspired by literature (at least science fiction literature) is the area of machines building machines. Countless sci-fi stories have been written about intelligent machines becoming smarter than humans and taking over the species. While software programs and powerful computer hardware have been developed that can beat all human comers in a chess match, the prospect of artificial intelligence taking control of our world has taken a back seat to other fears. However, the new research into machine design and building machines offers both hope and fear. In the science fiction stories, the highly intelligent computers were usually defeated by humans who figured out how to turn off the electrical power. However, if machines become smarter and they can also procreate, then there is another population bomb to worry about.

If a machine is to design a machine that it intends to build, it should also design the manufacturing process by which it will build the new machine. While the concept of machines becoming totally self-reliant for procreation requires an intense imagination, the concept of a machine designing other machines and the appropriate manufacturing processes that go with them is within the realm of current computer-aided design and engineering and simulation technology.

There are phenomena that may be speeding up the march to the reality of machines designing and building new machines. As the production lines of the e-factory become more cellular, more flexible, more dependent on modular product designs, and more computer controlled, the ability of computer-based systems to design manufacturing processes for modular products will increase. As the diversity of the universe of e-factory-oriented manufacturing cells expands, the number of potential production line configurations that could be assembled to produce different types of products expands. Thus, it is possible that at some point in the future, random or combinatorial methods may exist that would design a production process by configuring manufacturing cell designs. The best production process could then be selected on the basis of cost or speed.

This is already happening in the semiconductor industry. The newest semiconductor fabs — especially those called silicon foundries — operate as manufacturing nodes or cells with design rules by which they can build semiconductor products. In offices elsewhere in the world semiconductor-based machines (with some support by humans) design new semiconductor-based machines that are to be built in the new semiconductor fabs. The manufacturing rules of the fab are used to decide which design survives, i.e., are buildable in the fab.

Thus, it possible to see a future where there is a sea of e-factory nodes spread across the Internet-linked virtual world from which a machine intelligence could configure a supply chain of e-factories to produce a new product which no one company may have ever produced. The machine intelligence would have to ensure that the appropriate design information is created and distributed to the appropriate e-factory nodes. It would also have to ensure that the appropriate commercial agreements and pricing negotiations are created between the e-factory nodes and the appropriate commercial enterprises involved. The trends of e-markets, on-line negotiation, and collaborative product creation are all contributing to an environment that can support these requirements.

Summary

Creating the e-factory brings with it opportunities and risks. The opportunities include realizing the benefits of transformation of business processes and entire enterprises. The risks include the chances of choosing technology and designing processes that, when put together, do not achieve the performance levels needed. The roadmap methodology outlined herein is an organized approach to analyzing the technology and process needs of a manufacturing organization and identifying the technology solutions necessary to respond to those needs. The roadmap approach also makes it possible to deal with the rapid changes and evolution of the third wave of Internet-inspired technology for the e-factory.

Appendix

Solution Set Component Lists

As discussed in Chapter 6, trends in technology and processes have accelerated the introduction of new technologies into the e-factory. As this has happened, decision-making about which software and hardware suppliers to use become has become more complex. As the promise of total solutions from single-source enterprise suppliers gave way to the reality and necessity of using best-of-breed solution sets, the task of picking the best solution set has become a major task for e-factory planners.

Included in this appendix are lists of companies that have offered software that falls within the solution component categories discussed in Chapter 6. These lists were incomplete when they were compiled and are changing every month, but they can be used as starting points for building candidate solution sets. These lists are not exhaustive; they are only samples of the universe of companies that may offer software or services in these areas.

Call Center Management Systems

Austin Logistics	EWare	Magic
BisGen	IMAJ	Unica
Cincom	Interactive Software	Voxco
Cisco	IVRS	

Computerized Maintenance Management Software

ABC Technologies Int., Inc.	Advanced Maintenance Solns.	Allina Health System Q2000
ACT	Aero Systems Corp.	American Quality Sys., Inc.

American Services Resources

Angus Systems Group Ltd.

Aperture Technologies, Inc.

Argos Software

AssetWorks, Inc.

Barcontrol Sys. & Services

Beamex, Inc.

Benchmate Systems, Inc.

Bender Engineering, Inc.

Berland Technologies, Inc.

Black & Mcdonald Limited

Bonner & Moore Associates

Cadworks, Inc.

Canatech Consulting Int'l Ltd.

Caver-Morehead Systems

CHAMPS Software, Inc.

CK Systems, Inc

Compliance Tech., Inc.

Computron Software, Inc.

Control Pak International, Inc.

Cosys LLC

Creative Management Sys.

Cybermetrics

Datastream Systems, Inc.

Data-Trak, Inc.

Decision Dynamics, Inc.

Desktop Innovations, Inc.

DFM Systems, Inc.

Dima Litvak Corporation

DP Solutions, Inc.

DPC, Inc.

Dynacomp, Inc.

Eagle Technology, Inc.

ECRI

EDS

EFAX Corporation

E-Max, Inc.

Engineering Mgmt. Consult.

Epix, Inc.

EQ2, Inc.

Equipac Systems

Facility Management Tech.

FBO Systems, Inc.

Fields & Screens, Inc.

FKW Technologies Inc.

Fleming Systems

Fluor Daniel, Inc.

FMS Systems Corp.

Fortress Technologies

Four Rivers Software., Inc.

GBS Associates, Inc.

Generation Systems, Inc.

GP Solutions, Inc.

Hansen Information Tech.

HSB Reliability Technologies

IFS, Inc.

Indus International, Inc.

Innovative Tech Systems, Inc.

Integrated Software, Inc.

International Information Svcs

IQS, Inc.

ITX Stanley, Inc.

Ivara Corporation

J.D. Edwards Company

Johnson Controls, Inc.

Josalli, Inc.

Kakari Systems, Ltd.

Kenonic Controls Ltd.

Macola Software

MainSaver

Management Planning Syw.

Management Tech., Inc.

Marcam Solutions, Inc.

Marine Management Systems

MegaMation Systems, Inc.

Meridium

MicroMain Corporation

Microwest Software Sys., Inc.

Mincom, Inc.

Nielsen-PM Associates, Inc.

Norwich Technologies

Ogden Government Services

Omnicomp, Inc.

Operendum Software, Inc.

Ounce of Prevention Soft.

Owen Engineering

P.M. Sulcs & Associates Ltd.

Pearl Computer Systems, Inc.

Penguin Computer Consult.

Phoenix Data Systems, Inc.

PMS Systems Corp.

Precision Maintenance Sys.

Primavera Systems, Inc.

Prism Computer Corporation

Project Services Int., Inc.

Prototype Incorporated

PSDI

QBIC III Systems, Inc.

Rambow Enterprises

Ramco Systems Corporation

Raytheon Constructors, Inc.

Richard Morton Co.

Rippe & Kingston Systems
Roesel, Kent & Associates
Ross Systems, Inc.
SAP AG
Selfware, Inc.
Servidyne Systems, Inc.
Shoham Consulting
Sofwave, Inc.
Somax, Inc.
Spar Associates, Inc.
Spectec
SSA
SSI Services, Inc.
STN Computer Systems
Summit Software Sys., Inc.
Synergen Associates, Inc.
Syska & Hennessy
Tangible Vision, Inc.
Tecworks, Inc.
The Harrington Group, Inc.
The London Group, Inc.
The Omega Consultants
TMA Systems, Inc.
TMT Software Company
User Solutions Spreadsheet
Vitec, Inc.
Walker
Walsh Automation, Inc.
Western Software Solutions
WinterCress Dev., Inc.
Zonic Corporation

Customer Relationship Management Systems

Agillion
Allegis
Ardexus
Ascendant Solutions
Ascent Systems
Athene Software, Inc.
Broadway & Seymour Group
CallBack Software
Caspian
Cognitor, Inc.
Commence Corporation
ContinuitySolutions,Inc.
Epicor Software Corporation
eWare
Firstwave Technologies, Inc.
Foundation Network
Global Groupware Solutions
GoldMineSoftwareCorp
IBM
IMA
Infinium
Interface Software
Janna Systems
KnowledgeSync
LeadMaster
MarketForce
MarketProminence
Metrix
Million Handshakes
Nteration Systems, Inc.
Ockham Technologies
ON!contact Software Corp
Onyx Software
Partnerware, Inc.
Pivotal.com
Pricedex Software, Inc.
Rapport
Relavis Corporation
Remedy Corporation
SalesLogix
SalesManager Software
SAS Institute
Selligent
Silknet
Slingshot Software
Software Innovation
SPSS
SuperWiz
Symix
Synchrony Comm, Inc.
TASC
TeleMagic
Touchtone
TPS Labs
Unitract Technologies, Inc.
Update.com
v-TEAM
Webridge
Xpansions
YOUcentric Solutions

Electronic Procurement Systems

Ariba
Cobalt Group
Commerce One
Concur Technologies
DoveBid
ec-content
elcom.com
Epicentric
Extensity, Inc.
Firmbuy
PartMiner
RightWorks
Trilogy Software
VerticalNet
Vitadata

Enterprise Resource Planning Systems

Adonix
Advanced Software
  Designs
Asprova
Axis Computer Systems
Baan Company
Baeurer International
Cincom Systems
CMS Manufacturing
  Systems, Inc.
COSS Systems, Inc.
Eagle Technology
Epicor Software
  Corporation
Expandable Software,
  Inc.
HAL Systems

IFS Industrial and
  Financial Systems
IndustriOS Ltd.
Intentia International AB
  (publ)
Intuitive Manufacturing
  Systems
IQMS
J.D. Edwards Company
JUMP Technology
  Services
Lahey Software
Lilly Software Associates
Manufacturing Action
  Group, Inc.
Metasystems, Inc.
MISys, Inc.

Oracle
Pactus, LLC
PeopleSoft
Production Modeling
  Corporation
ProfitKey International.
Ramco Systems
  Corporation
ROI Systems, Inc.
SAP AG
ShopPro Software
STR Software Company
Symix
Synergen, Inc.
Telesis Software, Inc.
Think & Do Software, Inc.
TopTier Software

Factory Automation Systems

Abba Computer Systems
ACS Software, Inc.
Action Systems
  Associates, Inc.
Adaptable Business
  Systems, Inc.
Adept Technology, Inc.
AIM Computer
  Solutions, Inc.
American Advantech
  Corporation
Applied Statistics, Inc.
ASAP, Inc.
AutoSimulations, Inc.
B.R. Willoughby &
  Associates
Batch Process
  Technologies, Inc.
Blue Mountain Quality
  Resources
BMS, Inc.
Boothroyd Dewhurst, Inc.
Bradley Ward Systems, Inc.
Camstar Systems, Inc.

Canary Labs, Inc.
Casco Development, Inc.
Cetec Automation, Inc.
Chesapeake
Cimnet Systems, Inc.
CimWorks GageTalker
CIMx
CMS Research, Inc.
Coastal Data Products, Inc.
Competitive Strategies,
  Inc.
Computer Recognition
  Sys., Inc.
Concentus Technology
  Corp.
ControlSoft, Inc.
C-Pak Corporation
Data Collection Systems,
  Inc.
DataModes, Inc.
DataNet
DataTrack International,
  Inc.
Dataworks

Decision Dynamics, Inc.
DLog Remex, Inc.
DT Industries, Inc.
Eagle Technology, Inc.
EASE, Inc.
Enterprise Systems, Inc.
ExperTune, Inc.
Eyring Corporation
FactorySoft, Inc.
FactoryWare, Inc.
FasTrak SoftWorks, Inc.
Finley System, Inc.
Fluid Air, Inc.
Focused Approach, Inc.
Four Rivers Software
  Systems
GE Fanuc Automation
GP Solutions, Inc.
Greco Systems
Greycon
Hanford Bay Associates,
  Ltd.
Hartlepool Systems
  International

Hertzler Systems, Inc.
Hewlett-Packard Company
HMS Software, Inc.
i2 Technologies
ICC/GR Software
InFiSy Systems, Inc.
Instant Data Systems
Interpec
Interval Logic Corporation
Intuitive Data Solutions
Iota Development
    Corporation
IQS, Inc.
ITC Integrated Systems,
    Inc.
JC-I-T Institute of
    Technology
JOBSCOPE Corporation
JobTime Systems, Inc.
John A. Keane &
    Associates, Inc.
Kessler-Rollins, Inc.
LABTECH
LabVANTAGE Solutions
Lanner Group, Inc.
Litton PRC Corp.
Manufacturing Action
    Group, Inc.
Manufacturing Mgmt.
    Sys., Inc.
Manugistics, Inc.
MDSS
MDT Software
MES Solutions
    Incorporated
Metrscope International
Minitab, Inc.
Mission Critical Software

Mitron Corporation
Murphy Software
Mystic Management
    Systems, Inc.
Northwest Analytical, Inc.
Oncuity, Inc.
ONLINE Software Labs
ONSPEC Automation
    Solutions
Pacific Coast
    Engineering, Inc.
Panasonic Factory
    Automation Co.
Performance Software
    Associates
Pister Group, Inc.
POMS Corporation
PQ Systems, Inc.
Praedictus Corp.
PremierAutomation,Inc.
Primus Technology, Inc.
Process Integrity, Inc.
ProcesSoft Corporation
ProfitKey International
PROMODELorporation
Q-CIM, Inc.
Quality America, Inc.
Quality Measurement
    Sys. Corp.
Real World Technology
    Corp.
Realtime Information
    Systems
Reveille Technology, Inc.
Rockwell Software
Signature Technologies,
    Inc.
Simple Solutions, Inc.

Software Technology
Solumina
SSI Consulting Group
Starwin Industries, Inc.
Stat Soft, Inc.
Stat-Ease, Inc.
STG
Stochos, Inc.
Symix/Pritsker Division
Synetcom Digital, Inc.
Sysmark Information
    Sys., Inc.
TA Engineering
    Company, Inc.
TAL Technologies, Inc.
Talarian Corporation
Tangible Vision, Inc.
Taylor Manufacturing
    Systems
Techview Corporation
Tetra International, Inc.
Thedra Technologies
Think and Do Software,
    Inc.
Thru-Put Technologies
TIW Technology, Inc.
Trihedral Engineering Ltd.
UNIK Associates
Unlimited Solutions, Inc.
User Solutions, Inc.
Valstar Systems Ltd.
Verax Systems, Inc.
Vertex Industries, Inc.
VIA Information Tools
Waterloo Manufacturing
    Software
WeighAhead Systems
    PTY, Ltd.

Factory Floor Control Systems

ABB Ltd.
ACS Software, Inc.
Adept Technology, Inc.
Applix, Inc.

Batch Process
    Technologies, Inc.
Bradley Ward Systems,
    Inc.

CACI Products Company
Camstar Systems, Inc.
CascoDevelopment,Inc.
CIMLINC Incorporated

CIMx
Coastal Data Products, Inc.
Consilium, Inc.
Domain Manufacturing Corporation
DP Solutions, Inc.
DT Industries, Inc.
Eagle Technology, Inc.
Effective Management Systems, Inc.
Eyring Corporation
FactorySoft, Inc.
FactoryWare, Inc.
FasTrak SoftWorks, Inc.
Fluid Air, Inc.
Four Rivers Software Systems, Inc.
Fourth Shift Corporation
GE Fanuc Automation
Groupe Schneider
Hewlett-Packard Company
Hilco Technologies, Inc.
HK Systems, Inc.

HMS Software, Inc.
Honeywell, Inc.
IBM Corporation
Intellution, Inc.
Intercim
Invensys
JC-I-T Institute of Technology
Lanner Group, Inc.
Lilly Software Associates, Inc.
Logica Carnegie Group, Inc.
Marcam Solutions, Inc.
Mitron Corporation
Oncuity, Inc.
Panasonic Factory Automation Company
PowerWay, Inc.
Praedictus Corp.
PremierAutomation,Inc.
ProfitKey International
Promis Systems Corporation

Real World Technology Corporation
Reveille Technology, Inc.
Rockwell Automation
Simulation Sciences, Inc.
Steeplechase Software, Inc.
STG
Symix Systems, Inc.
SynQuest, Inc.
Sysmark Information Systems, Inc.
Taylor Manufacturing Systems
Thedra Technologies
Thru-Put Technologies
TIW Technology, Inc.
USDATA Corporation
Valstar Systems Ltd.
VIA Information Tools
Walsh Automation, Inc.
Wonderware Corporation

Logistics Management Systems

Catalyst International, Inc.
EXE Technologies, Inc.
J.D. Edwards

Manhattan Associates, Inc.
Manugistics, Inc.
McHugh Software International

Optum, Inc.
TRW

Manufacturing Execution Systems

Adept Technology, Inc.
Applied Materials (Consilium)
Aspen Technologies, Inc.
Baan
Bradley Ward Systems, Inc.
Brooks Automation (FASTech)
CACI Products Company
Camstar Systems, Inc.

Chesapeake Technologies
CIMLINC Incorporated
Cimnet Systems, Inc.
Consilium, Inc.
Eagle Technology, Inc.
Effective Management Systems, Inc.
Emerson Electric (Intellution)
Eyring Corporation
FactorySoft, Inc.

FactoryWare, Inc.
Four Rivers Software Systems, Inc.
GE Fanuc Automation
Gensym Corporation
Groupe Schneider
Hewlett-Packard Company, HP AIS
Hilco Technologies
HK Systems, Inc.
iBASEt

IBM Corporation
Intellution, Inc.
InterCim
Invensys (WonderWare)
Litton PRC Corp.
Manufacturing Action
   Group, Inc.
Marcam Solutions, Inc.
MES Solutions
   Incorporated
Mesabi Control
   Engineering: Mix Vision

Numetrix Limited
ORSI
POMS, Inc.
ProfitKey International
Promis Systems
   Corporation
Real World Technology
   Corporation
Rockwell Software
Siemens Industrial
   Automation
STG

Symix Systems, Inc.
Tangible Vision, Inc.
Taylor Manufacturing
   Systems
TIW Technology, Inc.
USDATA Corporation
ValuePak Solutions
Walsh Automation, Inc.
Wonderware
   Corporation

Product Data Management Systems

ACCESS Corporation
ACS Software, Inc.
Adaptive Media
Agile Software Corporation
Allegria Software
Altris Software, Inc.
American Software
Applied Automation
Auto-trol Technology
BAAN
Bluestone Software
Ceimis Enterprises, Inc.
Celerity Solutions
Centric Software
Cimmetry Systems, Inc.
Cincom Systems
CoCreate
CYCO Software
Deltek Systems
Digital Paper
Docucon, Inc.
Documentum, Inc.
eChange, Inc.
EDS
Eigner + Partner
Ensodex, Inc.
eQuorum
ESPS, Inc.
FFM Software, Inc.
FileNet Corporation
Foresight Software

Hagerman & Company,
   Inc.
IBM
IDFM
IDS Scheer
ImageMAX, Inc.
Immersive Design
Informative Graphics
   Corp.
Innodata Corp.
Integrated Support
   Systems
InterData Access
   Corporation
Intergraph Corporation
International Computex
International
   TechneGroup
ITI
J.D. Edwards
JBA International
LargeDoc Solutions
Layton Graphics
Manugistics
Marcam Solutions
MatrixOne
Metaphase Technology
O'Neil & Associates, Inc.
Oracle Systems
PC DOCS/Fulcrum
PeopleSoft

Platinum Software
PLP Digital Systems
PowerCerv Corporation
Powerway, Inc.
PSI Penta
PTC
QuadRite, Inc.
Quickstream Software
Rand Technologies
ROI Systems
SAP America
Scanning America
SDRC
Smooth Solutions, Inc.
Spicer Corporation
Staffware Corporation
Third Millenium
   Technology
TSA/ADVET
Unigraphics Solutions
Universal Systems
Vidar Systems Co.
WAMware Engineering
   Soln.
Warren-Forethought
Wilbanks Technology
Workgroup Technology
WorkPlace Systems
Xtend, Inc.

Sales Force Automation Systems

Abalon
Allegis
Ascent
Axcess Software
BisGen
Business Objects
CallBack
CASPIAN
ChannelWave Software
CHARTER continuum
Chordiant Software
Coral Sea Software

Datad Code
DataGlider
EN5
eSupportNow
eWare Ltd.
GoldMine Software Corp.
IMA
Infinium
Interact Commerce
Interactive Software
LeadMaster
Magic

salesforce.com
SalesLogix
SalesTactics.com
Slingshot Software
Software Innovation
SuperWiz
TASC
The CURA Group
Touchtone
Tranzline
Vikara
Xpansions

Supply Chain Execution Systems

Catalyst International, Inc.
Descartes Systems, Inc.
EXE Technologies, Inc.
HK Systems, Inc.
Industri-Matematik International Corp

Interactive Business Systems AB
J.D. Edwards
Manhattan Associates, Inc.
McHugh Software International

Optum, Inc.
Provia Software, Inc.
STS, Inc.
Swisslog Management AG
TRW
Vastera

Supply Chain Integration Software

Applix
BroadVision
CHARTER continuum
Coral Sea Software
CrossWorlds
Extricity

FileNET
IMAJ
Maconomy International Software
NCR
PeopleSoft

QAS
Saratoga Systems
Silknet
Sterling Software

Supply Chain Planning Companies

Adonix Transcomm, Inc.
Autosimulations
Baan
BMC Software, Inc.
Cambar Software
Catalyst International, Inc.
Cincom Systems, Inc.

Competitive Solutions, Inc.
Dataworks Corporation
DSA-Software, Inc.
Effective Management Sys.
Foresight Software, Inc.
Fourth Shift Corporation

Friedman Corporation
Frontec AMT
GLOVIA International
Haushahn Systems & Eng.
HK Systems
i2 Technologies

IFS Inc.
IMI Industries
Intentia International
J.D. Edwards
JBA International
Lawson Software
Lilly Software Associates
Logility, Inc.
LPA Software, Inc.
Macola Software
Manugistics, Inc.
MAPICS, Inc.

MK Group
Numetrix
Openplus International
Optum, Inc.
Oracle
PeopleSoft
Pivotpoint, Inc.
PowerCerv Corporation
Primavera Systems, Inc.
Qad Inc.
ROI Systems, Inc.
Ross Systems, Inc.

SAS Institute, Inc.
SCT
Silvon Software, Inc.
St. Paul Software
Symix Systems, Inc.
Synquest
Syspro Group
Taylor Manufacturing
  Sys.
Tetra International
Thru-Put Technologies
Western Data Systems

Warehouse Management Systems

Catalyst International, Inc.
EXE Technologies, Inc.
JDA Software Group, Inc.

Manhattan Associates, Inc.
McHugh Software
  International

Optum, Inc.
TRW

About the Author

Alex N. Beavers, Jr., Ph.D., is the Chief Executive Officer of Thomson Industries, a privately held, middle market manufacturing company that makes motion control components and systems for the semiconductor, automotive, aerospace, medical, and consumer goods industries. Prior to Thomson Industries, Dr. Beavers spent ten years with PricewaterhouseCoopers L.L.P., where his duties included being the Managing Partner of Supply Chain Management Services in North and South America. In that position, he was responsible for all e-business supply chain management consulting, process consulting, and e-business systems integration services. His career also includes being President and CEO of ITP Systems, a software and systems integration company that specialized in manufacturing systems integration for semiconductor fabs; and President and CEO of Applicon Technologies, a computer-aided-design-and-manufacturing (CAD/CAM) systems company. Early in his career, Dr. Beavers was with General Electric where he was in charge of strategic planning for the GE Factory Automation Group and General Manager of the Robotics and Vision Systems Division. He has an M.B.A. from Boston University, a Ph.D. and M.S. in electrical engineering from the University of Houston, and a B.S. in electrical engineering from Vanderbilt University.

Index